Electronic Packaging,
Microelectronics,
and Interconnection
Dictionary

Electronic Packaging and Interconnection Series
Charles Harper, Series Advisor

Published Books

To order or receive additional information on these or any other McGraw-Hill titles, in the United States please call 1-800-822-8158. In other countries, please contact your local McGraw-Hill representative.

MH92

Electronic Packaging, Microelectronics, and Interconnection Dictionary

Charles A. Harper

Martin B. Miller

McGraw-Hill, Inc.

New York San Francisco Washington, D.C. Auckland Bogotá
Caracas Lisbon London Madrid Mexico City Milan
Montreal New Delhi San Juan Singapore
Sydney Tokyo Toronto

Library of Congress Cataloging-in-Publication Data

Harper, Charles A.
   Electronic packaging, microelectronics, and interconnection
dictionary / Charles A. Harper, Martin B. Miller.
      p.   cm. — (Electronic packaging and interconnection series)
Includes index.
   ISBN 0-07-026688-3
   1. Electronic packaging—Dictionaries. 2. Microelectronic
packaging—Dictionaries. 3. Electronic apparatus and appliances—
Design and construction—Dictionaries. I. Miller, Martin B.
(Martin Boniface). II. Title. III. Series.
TK7804.H34   1993
621.381'046'3—dc20

                                                        92-40712
                                                                CIP

1 2 3 4 5 6 7 8 9 0 DOC/DOC 9 9 8 7 6 5 4 3

ISBN   0-07-026688-3

*The sponsoring editor for this book was Daniel A. Gonneau, the editing
supervisor was Kimberly A. Goff, and the production supervisor was
Donald F. Schmidt.*

Printed and bound by R. R. Donnelley & Sons Company.

Preface

It has been said that one of the biggest problems in electronics packaging, microelectronics, and interconnections is the language barrier. The reason is that this group of technologies is engineered by professionals from many disciplines, such as electrical, mechanical, and materials, as well as from many functions, such as engineering, manufacturing, and quality. The technical languages across this spectrum are quite different. It is not easy, for instance, for an electrical engineer to understand the chemical aspects of materials, or conversely, for a materials engineer to understand the electrical aspects of microelectronics. Multidisciplinary terminology is difficult enough, but acronyms and symbols worsen the problem. While once a livable situation, today this language barrier problem is unacceptable. The interdependence of disciplines and functions in modern electronics systems demands some understanding of all the disciplines and functions to achieve optimized system performance.

This was the need which led to the preparation of this dictionary. This dictionary covers all of the disciplines and functions in the broad range of electronic packaging, microelectronic, and interconnection subject areas. Thus, this dictionary is a valuable reference source to anyone in those fields by allowing specialists in any area to quickly access the terminology in all of the other areas. In addition, this dictionary includes perhaps the most extensive available listing of acronyms, symbols, and abbreviations. Since this so-called "alphabet soup" is such a constant problem in the industry, this feature alone will make this dictionary invaluable for most bookshelves. No dictionary such as this has ever been published. It is hoped, therefore, that this dictionary will serve its readers well. Since even the most thorough effort in a task of this magnitude will still miss some useful terms, acronyms, and symbols, we would greatly appreciate suggestions from our readers regarding additions or modifications for future editions.

Charles A. Harper
Martin B. Miller

Acknowledgments

In an effort to achieve the best possible standardization in this dictionary for terms and definitions which have not to date been standardized in the broad electronics industry, important support was obtained from leaders in four major industry groups. McGraw-Hill and the authors would like to sincerely acknowledge and express their appreciation to these leaders and industry groups for their cooperation in this effort to provide a working guide. These important industry groups which have supplied some of their standard terms and definitions for including in this dictionary are: the JEDEC Solid State Products Engineering Council of the Electronic Industries Association (EIA), the Institute for Interconnecting and Packaging Electronic Circuits (IPC), the International Society for Hybrid Microelectronics (ISHM), and the International Institute of Connector and Interconnection Technology (IICIT), JEDEC Publication No. 99, "Glossary of Microelectronic Terms and Symbols," and JEDEC Publication No. 100, "Terms, Definitions, and Letter Symbols for Microcomputers and Memory Integrated Circuits."

Within these groups, special thanks and credit is hereby given to: Ingrid M. Taylor of EIA/JEDEC and Daniel I. Amey of DuPont Electronics and Chairman of the JEDEC JC-11 Design Requirements Committee; the members of the JEDEC JC-10 Committee on Terms, Definitions, and Symbols; Raymond E. Pritchard, Thomas J. Dammrich, Dieter W. Bergman, and Anthony W. Hilvers of IPC; Bernard S. Aronson of ISHM; and Martin G. Freedman and Donald M. Chambers of IICIT. Their kind support will help the industry immeasurably in the increasingly important requirements of bridging the language gap between technologies and simultaneously achieving improved industry standardization of terms and definitions.

Charles A. Harper
Martin B. Miller

Electronic Packaging, Microelectronics, and Interconnection Dictionary

Ablative Material Plastics or other materials whose surface layers decompose when heated, leaving a heat-resistant layer of charred material. The successive layers break away, exposing a new surface. These plastics are especially useful in applications such as the outer skins of spacecraft, which heat up to high temperatures on reentry into the earth's atmosphere.

Abrasion Resistance (1) The ability of a material to withstand mechanical actions such as abrading, scraping, and scrubbing. (2) A measure of the ability of a wire to resist surface wear.

Abrasion Stripper A mechanical or motorized device equipped with buffing wheels that scrape insulation from conductors. *Also called buffing stripper.*

Abrasion Tester A laboratory testing machine for determining the abrasion resistance of the insulation on wire and cable. It consists of buffing wheels that scrape the insulation from electrical conductors.

Abrasive Trimming Adjusting thick-film resistors to nominal values by notching the resistors with a finely graded stream of abrasive material, such as aluminum oxide, directly against the resistor surface.

Absorption (1) The retention of moisture by a substance. (2) The dissipation or transformation of energy as it passes through a medium, such as a loss of electromagnetic energy when radio waves travel through the atmosphere; a loss of acoustic energy when sound waves pass through a medium; or a loss of kinetic energy when a nuclear particle is released as it passes through a body of matter.

Accelerated Aging The rapidly induced deterioration of a material, system, or device in a relatively short time by increasing voltage, temperature, and so on above normal operating values. The results give predicted service life under normal conditions. *Also called accelerated life test. See also High Temperature Reverse Bias Test.*

Accelerated Life Test *See Accelerated Aging.*

Accelerated Stress Test (1) A test conducted at a higher stress level than in normal operation and at a shorter time for the purpose of producing a failure. (2) A test in which conditions are intensified so as to allow critical data to be obtained in a shorter time period. *See also Accelerated Aging.*

Acceleration The rate at which the velocity of a moving body changes.

Accelerator (1) A chemical additive used to speed up a chemical reaction or cure; often used interchangeably with the term *promoter*. An accelerator is frequently used along with a catalyst, hardener, or curing agent. *See also Curing Agent.* (2) A device that gives very high velocities to charged particles such as electrons or protons.

Acceptance Tests Trials deemed necessary to determine the acceptability of a product, as agreed to by both purchaser and vendor.

Acceptor An impurity that when added in sufficient quantity to an intrinsic semiconductor material or to an N-type extrinsic material, converts it into a P-type semiconductor material. *See also Intrinsic Semiconductor.*

Access Hole A hole drilled through successive layers of a multilayer printed wiring board to gain access to an area, such as a land or pad, on one of the inner layers.

Accessories An assortment of mechanical parts such as cable clamps and other hardware that make up the complete connector assembly and are attached to the connector to form a total connector configuration.

Accordion (1) A Z-shaped spring contact, so shaped to give the spring a high deflection without excessive stress. These contacts are used in some printed circuit connectors. (2) A retractable cable

with a series of equally spaced transversed folds.

Acetal A rigid thermoplastic material that can be molded or extruded. Acetals are noted for their high tensile and flexural strength, and their good solvent and moisture resistance.

Acid Copper Copper electrodeposited using a sulfate-sulfuric acid bath.

Acrylic A synthetic resin made from acrylic acid or a derivative thereof. Acrylics offer good high-voltage tracking resistance and excellent transparency properties.

Acrylonitrile-Butadiene-Styrene (ABS) A low-cost thermoplastic that exhibits dimensional stability, high impact resistance, excellent mechanical properties, and high surface hardness over a wide temperature range.

Activating Layer A layer of material that renders a nonconductive material receptive to electroless deposition. *See also Electroless Deposition.*

Activation A chemical treatment that conditions the surface of nonconductive materials for improved adhesion of electrodeposited materials by the removal of oxides and/or the addition of an adhesion promoting catalyst to the surface.

Activator The chemical material used in the activation process. *See also Activation.*

2

Active Area The internal area of an electronic package, usually a cavity, where the substrate is attached. *See also Substrate.*

Active Components Electronic devices such as transistors and diodes that can operate on an applied electrical signal so as to change its basic characteristics (i.e., rectification, amplification, switching). *See also Passive Components.*

Active Devices Discrete devices, such as diodes or transistors, or integrated circuit devices, such as analog or digital circuits, in monolithic or hybrid form.

Active Element An element of a circuit in which an electrical input signal is converted into an output signal by the nonlinear voltage/current relationships of a semiconductor device. *See also Active Components.*

Active Metal A metal that is very reactive or very high in electromotive force. Typical examples are iron and aluminum. *See also Electromotive Force.*

Active Network A network containing both active and passive elements.

Active Substrate A substrate, such as silicon, in which active elements are implemented within the substrate material. *See also Substrate.*

Active Trimming Adjusting an electrical circuit with the power "on" to achieve a required functional output. It usually involves the trimming of thick-film resistors. *See also Functional Trimming.*

Actuator A part of a switch that, when activated by an external force, causes the switch to open or close.

Addition Polymerization A chemical polymerization reaction in which simple molecules (monomers) are added to one another to form long chain molecules without forming by-products. *See also Condensation Polymerization.*

Additive A substance added to materials, usually to improve their properties. Prime examples are plasticizers, flame retardants, and fillers added to plastic resins.

Additive Process A process in which conductive patterns are formed by the selective electrolytic deposition of conductive material on an insulating base material. *See also Subtractive Process.*

Add-On Component Any type of discrete or integrated prepackaged or chip component that is attached to a film network to complete the circuit function. *Also called add-on device.*

Add-On Device *See Add-On Component.*

Adhere To cause two surfaces to be held together by adhesion.

Adherend A body that is held to another body by an adhesive.

3

Adhesion The state in which two surfaces are held together by interfacial forces consisting of valence forces or interlocking action, or both.

Adhesion Layer A metal layer that bonds a barrier metal to a metal land on the surface of an integrated circuit. *See also Barrier Metal.*

Adhesion Promotion A chemical process that conditions the surface of a plastic or plastic parts and provides excellent adhesion of subsequent metal depositions.

Adhesion Promotor A chemical that prepares bonding surfaces to improve bond strength. *See also Primer.*

Adhesive A substance used for firm attachment of two surfaces by exerting forces of attraction between molecules of mating surfaces in an adhesive bond.

Adhesive, Anaerobic An adhesive that sets only in the absence of air, such as being confined between plates or sheets.

Adhesive, Contact An adhesive that is apparently dry to the touch and will adhere to itself instantaneously upon contact. *Also called contact bond adhesive or dry bond adhesive.*

Adhesive Failure Rupture of an adhesive bond such that the separation appears to be at the adhesive-adherend interface.

Adhesive, Heat-Activated A dry adhesive film that is rendered tacky or fluid by application of heat or heat and pressure to the assembly.

Adhesive, Pressure-Sensitive A viscoelastic material that in solvent-free form remains permanently tacky. Such material will adhere instantaneously to most solid surfaces with the application of very slight pressure.

Adhesive, Room-Temperature-Setting An adhesive that sets in the temperature range of 20 to 30°C (68 to 86°F), in accordance with the limits for standard room temperature specified in ASTM Methods D-618, "Conditioning Plastics and Electrical Insulating Materials for Testing."

Adhesive, Solvent-Based An adhesive having a volatile organic liquid as a vehicle.

Adhesive, Solvent-Activated A dry adhesive film that is rendered tacky just prior to use by application of a solvent.

Adjacent Conductor A conductor located next to another conductor either in the same multiconductor layer or in an adjacent layer.

Admittance The unrestricted flow of alternating current in a circuit. It is the reciprocal of impedance. The symbol is Y or y.

Adsorbed Contaminant A contaminant attracted to the surface of a material and held to that surface in the form of a gas, vapor, or condensate.

Adsorption The adhesion of gas or liquid molecules to the surface of solids or liquids with which they are in contact.

Advanced Composites *See Organic Composites.*

Advanced Statistical Analysis Program (ASTAP) A simulated circuit analysis program that performs direct current (DC), time domain, and frequency domain simulations to predict operating tolerances by statistical analysis. Included is a transmission line analysis program.

Aging The change in properties of a material or a component with time under specific conditions such as varying the temperature, pressure, or humidity. The result is an improvement or a deterioration of properties.

Aging, Accelerated *See Accelerated Aging.*

Aging, Natural A change in a material or component under normal environmental conditions such as oxidation, reaction with sulphur and other chemicals, and formation of reaction layers between base material and coating materials.

Aging Steam An artificial, accelerated treatment to the outer surfaces of components and boards to determine whether they can withstand storage requirements.

Agitation A process involving motion of a medium such as shaking, stirring, or vibration of a liquid.

Air Gap The air space between conductors, lands, pads, and other elements on a printed wiring board.

Airless Spraying A high-pressure process in which the pressure is sufficiently high to atomize the liquid coating particles without air. *See also Spray Coating.*

Air Vent A small gap in a mold to avoid entrapping gases in the plastic part during the molding process.

Alcohols Compounds characterized by the fact that they contain the hydroxyl (-OH) group. They are valuable starting points for the manufacture of synthetic resins, synthetic rubbers, and plasticizers.

Aldehydes In general, volatile liquids with sharp, penetrating odors, slightly less soluble in water than are corresponding alcohols. The group (-CHO), which characterizes all aldehydes, contains the most active form of the carbonyl radical and makes the aldehydes important as organic synthetic agents. They are widely used in industry as chemical building blocks for inorganic synthesis.

Align To accurately position or locate circuits, images, or patterns of a photomask over one another or over etched or screened patterns.

Alignment Marks *See Registration Marks.*

5

Aliphatic Hydrocarbon *See Hydrocarbon.*

Alkyd A thermosetting resin with excellent electrical properties that is used for molding a wide variety of electrical components.

Alligator Clip A mechanical spring-loaded device shaped like the jaws of an alligator. It is used as a temporary electrical connection at the end of a test lead.

Allowables A statistically derived estimate of a mechanical property based on repeated tests. It is the value above which at least 99 percent of the population of values is expected to fall with a confidence of 95 percent. *See also B Allowables.*

Alloy (1) In metallurgy, a substance formed by combining two or more metals but not necessarily with a chemical bond. (2) In plastics, a blend of polymers, copolymers, or elastomers under controlled conditions.

Alloy 42 An iron-nickel alloy whose low thermal expansion makes it especially suitable for use as glass seal feedthrough or leadframe material in the packaging of discrete integrated circuits and hybrid integrated circuits. *See also Kovar.*

Alloy Junction A point at which one or more impurity metals are melted on a semiconductor wafer to form P or N regions. *Also called fused junction.*

Alpha Particle A small, electrically positively charged particle with two protons and two neutrons that are thrown off at very high speeds by radioactive materials. It is capable of generating electron/hole pairs in microelectronic devices, causing soft errors in some components. Alpha has the lowest penetration of the various emitted particles, and will be stopped after traversing through only a few centimeters of air or a very thin solid film. *See also Electromagnetic Radiation.*

Alumina The polycrystalline form of aluminum oxide, Al_2O_3. A high-quality ceramic compound used to make ceramic substrates and chip carriers. It can withstand continuous high temperatures and has a low dielectric loss over a wide temperature range.

Alumina Tape Substrates that are made by casting alumina slurries to a predetermined thickness. This green tape or soft substrate material is then cut to size, holes are punched, and the tape is subsequently fired. *See also Green Tape, Slurry.*

Aluminum A lightweight metal with good corrosion resistance and excellent electrical and thermal conductivity, used in semiconductor technology to form the interconnections between devices on a chip. The bulk form is also machined to fabricate housings and chassis for electronic equipment.

Aluminum Nitride   A high-thermal-conductivity ceramic used in electronic packaging.

Aluminum-Steel Conductor   A composite of aluminum wires surrounding steel wires.

Ambient   The surrounding environment coming into contact with the system or component in question.

Ambient Temperature   The temperature of the surrounding cooling medium, such as gas or liquid, that comes into contact with the heated parts of the apparatus.

Amino Plastics   A group of thermosetting plastics, including melamine and urea, that are used in molding compounds.

Amorphous   An atomic or molecular structure that does not have a regular, patterned, crystallike arrangement. *See also Crystalline.*

Amorphous Polymer   A noncrystalline plastic that has no sharp melting point and exhibits no known order or pattern of molecular distribution. *See also Crystalline Polymer.*

Anaerobic   Free of uncombined oxygen. *See also Adhesive, Anaerobic.*

Analog   Pertaining to representation by means of continuously variable physical quantities whose values are usually displayed by a needle or similar indicator and printed numbers on a panel. *See also Digital.*

Analog Circuit   A circuit that provides a continuous relationship between input and output rather than a discontinuous or switching condition.

Analog Computer   A device that measures voltages, linear lengths, resistances, and light intensities. It can be manipulated by the computer and can also solve problems by setting up equivalent electrical circuits. Analog computers have limited capability to reduce approximate solutions, whereas digital computers give exact solutions. *See also Digital Computer.*

Analog Integrated Circuit   A type of linear integrated circuit intended to be used so that the output is a continuous mathematical function of the input. An operational amplifier is an example. *See also Digital Integrated Circuit, Integrated Circuit.*

Analog Microcircuit   *See Analog Integrated Circuit.*

Analog-to-Digital Converter   A device that converts an analog (continuous variable) signal into a proportional digital signal (output).

Angle Connector   A device that joins two conductors, end to end, at a specific angle.

Angled Bond   A wedge wire bond in which the first and second bonds are not aligned. *See also Wedge Bond.*

7

Angle of Attack In screen printing, the angle between the face of the squeegee and the surface of the screen. *Also called attack angle.*

Angstrom Approximately 4×10^{-9} inch, or 4×10^{-8} centimeter, or 4×10^{-4} micron, or 4×10^{-10} meter. *See also Micron, Mil.*

Anhydrides Compounds derived from the removal of water from organic acids. In general, they boil at higher temperatures than the corresponding acids. They are particularly valuable for reactions with the alcohol groups found in cellulose, phenols, sugars, and vegetable oils, yielding esters that are less water soluble than the original alcohol.

Anion A negatively charged ion that migrates toward the anode during electrolysis. *See also Ion.*

Anisotropic Pertaining to a material whose electrical, thermal, or optical properties are different in different directions through the material. *See also Isotropic.*

Anneal To heat and cool at a slow or prescribed rate to relieve mechanical stresses in metals or plastics and also to stabilize thick-film resistor materials. *Also called heat treat.*

Annealed Wire Wire that, after the final drawdown, has been heated and slowly cooled to remove the effects of cold working.

Annular Conductor A number of stranded wires that are twisted in three reversed concentric layers around a core.

Annular Ring The circular strip of conductive material that completely surrounds a hole on a printed circuit board.

Anode (1) The positive ($+$) terminal in an electroplating solution toward which the negative (-) ions flow. The part to be plated is usually the anode. (2) The electrode from which the forward current flows within a semiconductor diode or thyristor. *See also Semiconductor Diode, Thyristor.*

Anodic Cleaning Electrolytic cleaning in which the workpiece is the anode.

Anodic Silver A bar of silver used in electroplating baths. It is the sacrificial anode during the plating process.

Anodization An electrochemical oxidation process used to change the value of thin-film resistors or prepare capacitor dielectrics.

Anodize To form or deposit an insulating oxide layer on a metal part by electrolytic oxidation.

Antenna *See Aperture.*

Antioxidant A chemical used in the formulation of plastics to prevent or slow down the oxidation of material exposed to the air.

Antirotation A connector design with keying or locking provisions to sufficiently control positive orientation.

Antistatic Agents Agents that, when added to a plastic molding material or applied to the surface of a molded object, minimize electrostatic discharges from developing.

Antistatic Foam A static-free cushioning material. In the packaging of microcircuits, it provides shunt protection for leads and protection against electrostatic discharge and mechanical shock during handling, transporting, shipping, and storage. The soft, low-density version is flexible and noncorrosive. The semirigid, high-density version is fire retardant and provides shielding against electromagnetic interference and radio frequency interference.

Anvil The part of a crimping die that positions and supports the terminal during crimping. *Also called nest.*

Aperture An opening or form whose size determines the amount of electromagnetic energy intercepted. An antenna can serve as an aperture.

Application-Specific Integrated Circuit (ASIC) An integrated circuit chip designed for a specific application or product.

Aqueous Water based, as opposed to organic liquid based. *See also Slurry.*

Aqueous Cleaning The cleaning of electronic assemblies, especially circuit board assemblies, using deionized water or water-soluble agents. *See also Deionized Water.*

Aramid A highly oriented organic material derived from a polyamide but incorporating an aromatic ring structure. DuPont's Kevlar is a high-modulus aramid fiber.

Architecture (1) The organizational structure of a computing system, mainly referring to the CPU or microprocessor, but also including other hardware and software. (2) The specification of the relationships among the parts of a computer system.

Arcing Time The period of current flow after circuit protector activation and subsequent to element melting or contact opening.

Arc-Over Voltage The minimum voltage required to cause an arc between electrodes separated by gas or liquid insulation.

Arc Resistance (1) The time required for an arc to establish a conductive path in a material. (2) The characteristic of insulation materials to resist carbonization (tracking) of the material surface between electrodes as a result of voltage breakdown.

Area Array TAB Tape-automated bonding (TAB) in which an array of pads is located along the edge as well as the inner surface area of a substrate and is addressed in the bonding scheme. It is practical on

9

complex dies and integrated circuits where peripheral pad pitch cannot be reduced and more I/Os must be accommodated. *See also Tape Automated Bonding.*

Arithmetic Logic Unit (ALU) A part of a computer that performs arithmetic, logic, and related operations. Along with memory and control, it is an essential microprocessor element.

Armored Cable An electrical cable equipped with a metal wrapping, usually for mechanical protection.

Aromatic An organic molecule built around the hexagonal, six carbon benzene structure. It is usually more thermally stable than linear or branched aliphatic structures. *See also Benzene Ring, Hydrocarbon.*

Aromatic Hydrocarbon *See Hydrocarbon.*

Array A group of elements or circuits arranged in rows and columns on a substrate or printed wiring board.

Array Devices Devices without individual enclosures, each one having at least one electrode connected to a common conductor or all electrodes connected in series.

Artificial Intelligence The capacity of a machine to perform functions that are normally associated with human intelligence, such as reasoning and learning.

Artwork An accurately scaled configuration produced at an enlarged ratio to enable the product to be made therefrom by photographic reduction to a 1:1 working pattern. It is commonly used to create thick-film screens and thin film masks.

As-Fired (1) Pertaining to the smoothness of ceramic substrates. (2) The values of thick-film resistors as they emerge from the furnace before polishing or trimming.

Aspect Ratio (1) The ratio of the length of a film resistor to its width (L:W); equal to the number of squares of the resistor. (2) The ratio of the length or depth of a hole to its diameter (unplated) in a printed wiring board.

Asperites Depression or protrusions located on a smooth surface.

Assembly (1) A group of materials or parts, including adhesive, that have been placed together for bonding or have been bonded together. (2) A hybrid circuit to which discrete or integrated components have been attached. (3) A microelectronic device prior to packaging or encapsulation.

Assembly Drawing A drawing of all the parts and subassemblies mounted in their proper location and orientation to the circuits and interconnections to the film network.

Assembly Time The time interval between the spreading of an adhesive on the adherend and the application of pressure or heat, or both, to the assembly.

A Stage The initial, uncured stage in the curing of a thermosetting resin. *See also B Stage, C Stage.*

Atactic In hydrocarbons, the molecular structure of radical groups arranged heterogeneously around the carbon chain. *See also Hydrocarbon.*

Atomic Oxygen Resistance In space applications, the ability of plastics and elastomers to withstand atomic oxygen exposure.

Attach In hybrid assembly, to join permanently by means of intermediary material(s). Examples are attaching a substrate to a case and joining a device to a substrate.

Attack Angle *See Angle of Attack.*

Attenuation (1) The decrease in amplitude of a signal from one point to another in an electrical system. (2) The power loss, in decibels, in a length of cable. (3) The lessening of light intensity resulting from absorption and scattering of the light as it travels through optic fibers.

Audio Pertaining to the function of detecting, transmitting, or processing signals that have frequencies within the range sensed by the human ear. These frequencies range from 30 to 15,000 Hz.

Au/Ge An alloy of gold and germanium usually in an 88:12 ratio. It is used for higher-temperature die attachment applications.

Au/Sn A solder alloy of gold and tin, usually in an 80:20, ratio used for sealing gold-plated packages.

Autoclave (1) A closed vessel for conducting a chemical reaction or other operation under high pressure and heat. (2) In low-pressure laminating, a round or cylindrical vessel in which heat and pressure can be applied to a semi-impregnated paper or fabric positioned in layers over a mold.

Autoclave Molding After lay-up, placement of the entire plastic assembly in a steam autoclave at 50 to 100 lb/in^2. Additional pressure achieves higher reinforcement loadings and improved removal of air. *See also Lay-Up.*

Autodoping In semiconductor manufacture, the introduction of impurities from the substrate into the epitaxial layer during the process of epitaxy. *See also Epitaxy.*

Automated A combination of mechanical and electrical techniques and facilities programmed to perform a function without human effort.

Automated Component Insertion The act or operation of assembling discrete components to printed circuit boards by means of electrically controlled equipment.

Automatic Component Placement Software that automatically optimizes the layout of components on a printed circuit board.

Automatic DIP Insertion A mechanical technique in which a dual in-line package (DIP) is machine-inserted into proper placement. The package height is controlled at a predetermined distance above the printed wiring board by shoulders on the leads.

Automatic Wire Bonding Fine wire bonding of dies from die pads to substrate lands or headers, at a high rate of production with critical parameters preprogrammed for accuracy and reliability. Automatic wire bonding is accomplished by high-throughput machines with built-in microprocessors and other features. The program control software package addresses all bond parameters, including bond location, die height, bonding force and time, wire loop shape and height, tail length, reset height, ball size, ultrasonic power variables, electronic flame-off, and self-diagnosis. The pattern recognition video image received is digitized and usually compared with a target image stored in the machine's computer memory. (Video pixels are converted into binary information regarding bond details.) When

the two images correspond, the machine recognizes the proper orientation and is ready to perform automatic wire bonding.

Average Molecular Weight The molecular weight of the most typical chain in a given plastic. There will always be a distribution of chain and, hence, molecular weights in any polymer.

Average Wire Length The average length of wire of all the connections in a specific package level.

Axial Displacement An incremental difference between the initial position and final position after a force is applied along the components axis.

Axial Lead A wire coming out from a discrete component or device along the central axis of the device rather than from the sides.

Axial Strain Young's modulus of elasticity indicating strain applied along the axis of a material.

Azeotropic Mixture A mixture of several polar and nonpolar solvents to remove both polar and nonpolar contaminates. The mixture has a constant boiling point and cannot be separated by distillation.

B Allowables A statistically derived process of a mechanical property based on tests. The value above which at least 90 percent of the population of values is expected to fall with a confidence of 95 percent. *See also Allowables.*

Back Bonding Joining integrated circuits to a substrate by applying adhesive to the back side of the chips. *Also called back mounting.*

Back End of the Line (BEOL) Integrated circuit fabrication processes such as dicing the wafer into individual integrated circuits, wire bonding the transistors and resistors to the pads on the substrate, and making chip-to-package connections.

Backfilling (1) Filling an evacuated hybrid circuit package with a dry, inert gas prior to hermetically sealing. (2) Filling vias with a solid conductive ink in order to aid in electrically connecting two conductive layers. The ink solidifies after curing.

Back Mounted A connector that is mounted within a box/panel, with its mounting flanges located inside the equipment.

Back Mounting *See Back Bonding.*

Backplane *See Backplane Panel.*

Backplane Connector An interconnection assembly configuration having terminals, such as solderless wrapped terminals, on one side and connector receptacles on the opposite side. These are used to provide point-to-point electrical connections. *See also Mother Board.*

Backplane Panel An interconnection panel in which printed wiring assemblies or integrated circuit packages can be plugged or mounted.

Back Radius The radius of the trailing edge of a bonding tool foot.

Backshell Mold A mold that is used for molding the back of a connector after the cable connections have been installed.

Backside Metallurgy (BSM) A metallized pad that is electrically connected to internal conductors within a multilayer package, and to which pins are brazed.

Back Sputtering *See Sputter Cleaning.*

Bagging Applying an impermeable film over an uncured lay-up part and sealing the edges so that a vacuum can be drawn. *See also Lay-Up.*

Bag Molding The techniques of molding reinforced plastic composites using a flexible cover over the rigid mold. The composite material is placed with the bag or flexible cover. Pressure is then applied by drawing a vacuum or inflating the bag. *See also Bagging.*

Bake-Out An elevated temperature cycle that expels unwanted gases and moisture prior to sealing hybrid circuit electronic packages.

Balanced Amplifier (1) A one-output amplifier in which quiescent DC output voltage is reduced to zero or a specified level. (2) A two-output amplifier in which the difference between the quiescent DC output voltages is reduced to zero or a specified level. *See also Quiescent Voltage or Current.*

Ball Bond A type of wire bond in which an interconnection wire, flame-cut to produce a ball-shaped end, is deformed by a thermo-compression tool on a metallized pad. *Also called a nail head bond* (because of its flattened appearance).

Ball-Limiting Metallurgy (BLM) The metallurgy that controls the size and area of solder connections, limits the flow of solder balls to the desired area, and provides adhesion and contact to the chip wiring.

Bar Code A coding system consisting of vertical lines that can be read by an optical scanner and thereby converted into machine language.

Barcol Hardness A value obtained using a Barcol hardness tester, which gauges hardness of soft materials, such as plastic, by indentation of a sharp steel point under a spring load. *See also Indentation Hardness.*

Bare Board A printed wiring board that has been sheared to size, printed, etched, plated, and drilled but has no components mounted on its surfaces.

Barium Titanate A chemical compound, $BaTiO_3$, used in making high-dielectric-constant ceramic capacitors and high-dielectric thick-film pastes.

Barrel The crimped part of a terminal or contact. If it is designed to receive a conductor, it is called a *wire barrel*; if it is designed to support or grip insulation, it is called an *insulation barrel*.

Barrel Plating A manufacturing method in which a large number of small parts are placed in a rotating drum and plated.

Barrier A dielectric material that is used to insulate electrical circuits from one another or from ground.

Barrier Layer A thin metal coating that prevents or decreases the growth of intermetallics. For example, a barrier layer of nickel between copper and tin will decrease the growth of tin-copper intermetallics.

Barrier Metal A metal used to isolate the semiconductor die lands from the top metallization layer. *See also Adhesion Layer.*

Barrier Strip A continuous line of dielectric material used to insulate circuits from one another and from ground.

Base (1) The bottom of an electronic package, usually made of a high thermally conductive material. (2) An insulating support for printed patterns. (3) An alkaline chemical. *See also pH.*

Base Material The insulating material that supports the conductive patterns and electronic devices of a printed wiring board. It can be rigid or flexible.

Base Metal The metal from used to make connectors, contacts, and other accessory parts that are later electroplated.

Basic Device The simplest useful device that exhibits solid-state phenomena.

Basket Weave A type of weave in which two or more warp threads cross alternately with two or more filling threads. The basket weave is less stable than the plain weave but produces a flatter and stronger fabric. It is also more pliable than the plain weave and maintains a certain degree of porosity without too much sleaziness, though not to the extent of the plain weave. *See also Crow Foot Weave, Plain Weave.*

Batch Processing A manufacturing method in which large numbers of components are processed simultaneously.

Baud A unit of data transmission or signaling speed equal to the number of discrete signal events per second. It is equivalent to the bit rate in bits per second.

Bayonet Coupling A quick-coupling device for plug and receptacle connectors, accomplished by the rotation of a cam-operating device that brings the connector halves together.

Beam Lead A long, usually flat, metallic lead that extends beyond the body of a device. One end is attached to the chip device and the other is connected to circuitry on the substrate, providing an electrical connection or mechanical support, or both.

Beam Lead Chip A chip employing electrical terminations in the form of tabs extending beyond the edge of the chip for direct bonding to a mounting substrate.

Beam Lead Device An active or passive chip device having beam leads as its primary interconnection and mechanical attachment means to a substrate.

Beam Lead Tape A flat, flexible material, made of thin polyamide film and copper foil, that is formed into beam leads for tape-automated bonding. *See also Tape-Automated Bonding.*

Bed of Nails An array of probes in a predetermined grid pattern for contacting test points on a printed board or substrate for detailed

15

testing purposes. *See also Microprobe.*

Belled Mouth A flared or widened entrance to a connector barrel that provides easier insertion of the conductor or mating contacts.

Bellows Contact A connector contact with a flat spring that is folded to provide a uniform spring loading to the mating part.

Benzene Ring The basic structure of benzene, which is a hexagonal, six carbon atom structure with three double bonds. This is also a basic structure in organic chemistry. Aromatic structures usually yield more thermally stable plastics than do aliphatic structures. *See also Aromatic, Hydrocarbon.*

Beryllia Beryllium oxide (BeO). A ceramic material with high thermal conductivity that is used for thick-film substrates or for ceramic packages in high-power applications. *See also Aluminum Nitride.*

Beryllium Copper A metal with superior fatigue properties at high temperatures. Used in applications requiring repeated insertion and extraction.

Betascope A beta particle instrument that uses radioactive isotopes to measure plating and film thicknesses.

Bias Voltage The base voltage that establishes a semiconductor's desired DC operating voltage.

Bifurcated Slotted lengthwise, as with a flat spring, to provide additional points of contact.

Bifurcated Connector A hermaphroditic connector containing forklike mating contacts.

Bifurcated Contact A type of contact created when connectors made from a flat spring-type material are slotted lengthwise. The slotting provides more points of contact.

Binary A numbering system composed only of two digits, 1 and 0, used for digital integrated circuits.

Binary Digit (BIT) The smallest part of information in a binary system. A bit is either a 1 or a 0.

Binders Materials that are added to compositions to provide strength for handling purposes, such as binders added to thick-film pastes and unfired substrates prior to prefiring. They can be organic or inorganic materials and volatize during firing.

Binding Post A fixed support to which wire conductors are connected.

Bipolar and Complementary Metal-Oxide Semiconductor (BiCMOS) An integrated circuit process technology that combines bipolar and metal transistor types.

Bipolar Device A current-driven electronic device with two poles.

16

Bipolar Field Effect Transistor (BIFET) A linear circuit that combines bipolar and field-effect transistors (FETs) on the same chip, for FETs improved performance and cost savings.

Bipolar Transistor A transistor that uses both negative and positive charge carriers.

Birdcage A defect in a stranded wire. Specifically, a separation from the normal lay of the strands between the covering of an insulated wire and a soldered connection, such as an end-tinned lead.

Bis Epoxy *See Difunctional Epoxy.*

Bismaleimide A type of polyamide that cures by addition rather than by a condensation reaction. It generally has higher temperature resistance than epoxy.

Bit Rate The speed at which bits are transmitted, usually expressed in bits per second. *See also Binary Digit.*

Blackbody A radiator and absorber of radiant energy.

Black Box A loosely used term to describe an electronic device or assembly chassis within a larger system that can be mounted as a single package.

Black Oxide (1) A surface material formed by an oxidizing process onto mating surfaces of a metal package and used for improved sealing properties. (2) A single

material formed by a similar process onto the back surface of copper foil to provide improved bond strength of the foil onto a copper-clad laminate.

Blade Contact A flat male contact designed to mate with a flat female contact or a tuning fork, used in multiple-contact connectors.

Bleed (1) To give up color when in contact with a liquid medium. (2) In plastics technology, the surfacing of undesired materials.

Bleeding (1) A condition in which a plated hole discharges process material or solution from crevices or voids. (2) The lateral spreading or diffusion of a thick-film paste into adjacent unwanted areas beyond the geometric dimensions of the printing screen, as may occur during drying or firing.

Blending The mixing of different viscosities of the same types of thick-film pastes, conductive or resistive, to achieve intermediate viscosities or resistivities.

Blind Fastener A fastener designed for holding two rigid materials, with access limited to one side.

Blind Via A via that extends from one or more inner layers to the surface of a substrate or board. *See also Via.*

Blister (1) A localized swelling and separation between any of the layers of the base laminate or between the laminate and the metal cladding. It is a form of localized delamination. (2) A raised area on the surface of a molded part caused by the pressure of gases inside on its incompletely hardened surface. (3) In thick-film technology, a raised part of a deposited conductor or resistor formed by the outgassing of the binder or vehicle during the firing cycle.

Block (1) A connector housing. (2) In thick-film technology, to plug up the open mesh in a screen to prevent deposition in unwanted areas.

Block Copolymer A compound resulting from a chemical reaction between a number of molecules that are a block to one monomer and a number of molecules that are a block to a different monomer.

Block Diagram A circuit diagram in which the functional units of a system are shown in the form of blocks and the relationships among the blocks are indicated by connecting lines.

Blocking The ability of a semiconductor device or junction to offer high resistance to the passage of current.

Blowhole A hole or void in a solder connection caused by outgassing during soldering.

Blowing Agent A chemical added to a plastic in order to generate inert gases upon heating. This blowing or expansion causes the plastic to expand, thus forming a foam. *Also called foaming agent.*

Blow Molding A method of fabrication of thermoplastic materials in which a parison (hollow tube) is forced into the shape of the mold cavity by internal air pressure.

Boat A container, usually ceramic, designed to hold materials that are to be evaporated or fired. *See also Sputtering, Vacuum Deposition.*

Bobbin Lug A lug that serves to connect coil wires to external lead wires. The lug is usually mounted in a plastic or paper bobbin.

Boiling The changing of a liquid into a vapor accompanied by the formation of bubbles within the liquid.

Bolt-Type Connector A type of connector in which contact is made between the conductor and the connector by clamping bolts.

Bomb A chamber in which packages are stored for pressurizing or evacuating in the leak testing of electronic packages. *See also Fine Leak.*

Bond (1) The union of materials by adhesives. (2) To unite materials by means of an adhesive. (3) An attachment between a die and substrate or substrate and package using an adhesive for mechanical reasons or an interconnection such as a thermocompression or ultrasonic wire bond to perform an electrical function. *See also Wire Bond.*

Bondability The surface conditions, such as dryness and cleanliness of the bonding areas, that are necessary to provide a reliable bond.

Bond Deformation A change in shape of the lead made by the bonding tool, causing plastic flow in making the bond.

Bond Envelope The established range of bonding parameters over which acceptable bonds can be made. *See also Bond Schedule.*

Bonding The joining together of components for mechanical, electrical, or sealing functions.

Bonding Area A metallized area at the end of a thin metallic strip or on a semiconductor chip to which a wire bond or an electrical connection is made. *Also called bonding island, bond land, bond pad, or bond surface.*

Bonding, Die The attaching of a semiconductor chip to a bonding area on a substrate with a conductive or dielectric adhesive or a eutectic or solder alloy. *See also Die Bond.*

Bonding Island *See Bonding Area.*

Bonding Layer An adhesive layer used in bonding other discrete layers during lamination.

Bonding Pad An area on a chip to which a connection to the chip can be made. *Also called bonding island, bond site, or bond surface. See Bonding Area.*

Bonding Tool The instrument used to position leads or discrete wires over a land with sufficient energy to complete the termination. *See also Capillary.*

Bonding Wire Fine gold or aluminum wire for making electrical connections in hybrid circuits between various bonding pads on the semiconductor device substrate and device terminals or substrate lands.

Bond Interface The interface between the gold wire and the bonding area on the substrate.

Bond Land *See Bonding Area.*

Bond Lift-Off The failure mode whereby the bonded lead separates from the surface to which it was bonded. *Also called bond-off.*

Bond Line *See Glue Line.*

Bond-Off *See Bond Lift-Off.*

Bond Pad *See Bonding Area.*

Bond Parameters *See Bond Schedule.*

Bond Schedule The preestablished values of the bonding machine parameters used when adjusting for bonding. Some important values are bonding pressure, bonding time, and bonding temperature.

Bond Separation The distance between the attachment points of the initial and second bonds of a wire bond.

Bond Site The part of the bonding area where the wire bond was made.

Bond Strength (1) The unit load applied in tension, compression, flexure, peel, impact, cleavage, or shear, required to break an adhesive assembly with failure occurring in or near the plane of the bond. (2) In wire bonding, the strength at the breakpoint of the bond interface, measured in grams. (3) The force per unit area required to separate two adjacent layers by a force perpendicular to the board surface; usually refers to the interface between copper and base material.

Bond Surface *See Bonding Area.*

Bond-to-Bond Distance The distance, measured from the bond site on the die to the bond impression on the post, substrate land, or fingers, that must be bridged by a bonding wire or ribbon.

Bond-to-Chip Distance In beam lead bonding, the distance from the heel of the bond to the component.

Boot A mold placed around the wire terminations of a multiple-contact connector to contain the liquid potting resin during cure.

Boresight The forward direction on the axis of a parabolic dish antenna. In practical terms, the direction in which an antenna is pointing.

Boron Fibers High-modulus fibers of elemental boron vapor deposited onto a thin tungsten wire to impart strength and stiffness. Supplied as single strands or tapes.

Borosilicate Glass A type of glass having a closely matched coefficient of expansion between metal leads (Kovar) and ceramic packages. *See also Glass-to-Metal Seal.*

Boss A projection on a plastic or metal part designed to add strength, facilitate alignment during assembly, and permit attachment to another part.

Bottom Metallization The metallization on the bottom surface of a chip to permit bonding it to a mounting substrate. *See also Metallization.*

Boule An ingot of semiconductive material, usually 6 to 8 inches in diameter, from which slices are cut and processed into wafers. Semiconductor devices are then fabricated on these wafers. *See also Semiconductor Device, Wafer.*

Bow The deviation of a base substrate material or printed circuit board from flatness.

Box Pattern A pin arrangement for plug-in packages in which the pins are arranged in rows, forming a square or rectangle.

Box-Style Wire Contact A design feature in which the wire is completely enclosed in a contact and cannot be pushed through the connector.

Braid (1) A woven metallic wire used for shielding or as a ground wire. (2) A woven, fibrous protective outer covering over multiple conductors, as in a cable.

Branch Connector A connector that joins a branch conductor to the main conductor at a specific angle.

Brass An alloy of copper and zinc with optional small amounts of tin and lead. Its low cost and excellent electrical conductivity make it a good candidate for electrical applications. Brass reaches its yield point at a low deflection force; hence, it deforms easily and fatigues slowly.

Brassboard An arrangement of components interconnected in a field-demonstrable assembly. *See also Breadboard, Prototype.*

Braze To join metals with a nonferrous alloy at temperatures above 800°F. *Also known as hard soldering.*

Brazed Terminal A terminal with a barrel seam brazed to form one piece.

Brazing The joining of two similar or dissimilar metals by melting a non-ferrous metal or alloy having a lower melting point than the base metals.

Breadboard An arrangement of components wherein the components are temporarily interconnected to determine the feasibility of some circuit. This constitutes a laboratory-demonstrable circuit. *See also Brassboard, Prototype.*

Breakaway In screen printing, the distance between the top surface of the substrate and the underside of the screen when the squeegee is not in contact with the screen.

Breakdown The degradation of properties of a dielectric or insulating material that results in electrical failure.

Breakdown Voltage The voltage at which an insulator or dielectric ruptures or at which ionization and conduction take place in a gas or vapor. *See also Dielectric-Withstanding Voltage.*

Breakout The location in a multiconductor cable at which one or more conductors are separated from the cable to complete circuits to other points.

Bridging, Electrical (1) The formation of a conductive path between conductors. (2) A defective condition in which the localized separation between any two conductor lines or paths is reduced to a point determined by some specified electrical test value. Bridging is caused by misalignment, screening, solder splash, smears, or foreign material.

Bridging, Solder The filling of the space between parallel conductors, which are close together, with solder. Bridging can occur on the surface of printed wiring boards, downstream from where leads protrude from leaded packages.

21

British Thermal Unit (BTU) The quantity of heat required to raise the temperature of one pound of water through one degree Fahrenheit. This quantity depends on the absolute temperature of the water. One mean BTU is equal to 2326 joules per kilogram.

Brushes Conductors, normally of some carbon or graphite composition, that are used to achieve electrical connection between stationary and moving parts of electrical machines such as motors or generators. *See also Slip Rings.*

B Stage An intermediate stage in the curing of a thermosetting resin. B-stage resin can be heated and caused to flow, thereby allowing final curing in the desired shape. *Also called prepreg*, for fibers impregnated with B stage resin. *See also A Stage, C Stage.*

B-Tab *See Bumped TAB.*

BT Laminate A laminate containing a mixture of bismaleimide triazine resins that exhibits higher thermal stability than FR-4 epoxy-glass laminates. *See also Cyanate Ester, Difunctional Epoxy, and Non-Amine-Cured Epoxy.*

Bubble Test *See Gross Leak Test.*

Buffer (1) A protective overlay, normally of resilient material, surrounding a coated fiber to provide additional mechanical isolation and protection. (2) A circuit employed to minimize the effects of a following circuit on the preceding circuit.

Buffing Stripper A tool that removes flat cable insulation from conductors. *Also called abrasion stripper.*

Build The increase in thickness of a metal conductor, especially magnet wire, because of insulation.

Built-In Test (BIT) Circuits included in operational equipment expressly for the purpose of providing for on-line automatic testing.

Bulk-Channel Charge-Coupled Device *See Buried-Channel Charge-Coupled Device.*

Bulk Conductance Conductance between two points of a homogeneous material.

Bulk Density The density of a molding material in granular, nodular, or other loose form expressed as a ratio of weight to volume (e.g., g/cm^3 or lb/ft^3).

Bulk Factor The ratio of the volume of loose molding powder to the volume of the same weight of resin after molding.

Bulkhead Connector A connector designed to be inserted in a panel cut out from the component side.

Bulk Resistance That portion of contact resistance attributed to the length, cross section, and type of material. *See also Contact Resistance.*

Bulk Rope Molding Compound Molding compound made with thickened polyester resin and fibers less than

½ inch long. Supplied as rope, it molds with excellent flow and surface appearance.

Bump A small metal mound or hump that is formed on the chip or the substrate bonding pad and is used as a contact in face-bonding. It is a means of providing connections to terminal areas of a device.

Bump Contacts (1) Raised areas on a device, such as the base, emitter, and collector of a transistor chip, that are used for alloying with pads on a substrate during the mounting and interconnecting operations. (2) Contacting pads that rise substantially above the surface level of the chip. (3) Raised pads on the substrate that contact the flat land areas of the chip.

Bumped Chip A chip that has on its termination pads a bump of solder or gold used for bonding to external · contacts. This allows for bonding of all leads simultaneously as opposed to one at a time, as in wire bonding. *See also Gang Bonding.*

Bumped TAB Tape-automated bonding (TAB) wherein the raised solder bump is attached to the tape material instead of the bump being on the chip. *See also Tape-Automated Bonding.*

Bumped Tape *See Bumped TAB.*

Buried Channel A transfer channel beneath the surface of the semiconductor. *See also Channel.*

Buried-Channel Charge-Coupled Device (BCCD) A charge-coupled device that confines the flow of charges to a channel lying beneath the surface of the semiconductor. *See also Charge-Coupled Device.*

Buried Layer (1) A distinguishable region introduced under a semiconductor circuit element, such as under the collector region of a transistor, to reduce the series collector resistance. (2) In multilayer printed wiring boards, the inner conductor layers.

Buried Resistors Thick film resistors that are located on inner layers of multilayer substrates in order to reduce conductor lengths.

Buried Via A via that connects inner layers but does not extend to the surface of a substrate or board. *Also called hidden via.* See Fig. 1.

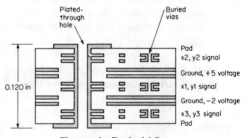

Figure 1: Buried Via

Burn-In Subjecting electronic components to elevated temperatures, normally 125°C, for an extended period of time, such as a commonly used standard of 168 hours, to stabilize their characteristics or cause failure to marginal devices.

23

See also Bias Voltage, High Temperature Reverse Bias Test.

Burn-In, Dynamic Burn-in, under actual operating conditions.

Burn-In, Static Burn-in, under a constant voltage rather than actual operating conditions, either forward or reverse bias.

Burning Rate The tendency of plastic materials to burn at given temperatures.

Burn-Off Cutting a bonding wire by passing a flame across it, thereby melting the wire. Used in gold wire thermocompression bonding to form a ball in the subsequent forming of a ball bond. *Also called flame-off.*

Bus A conductor or assembly of conductors used for transmitting signals or power from one or more sources to one or more destinations. *See also Control Bus, Data Bus.*

Bus Bar A heavy copper or aluminum strip or bar used to carry large amounts of current.

Bus Reactor A current-limiting device that is connected between two busses, or between two sections of one bus, to limit any disturbances caused by either the bus or bus section.

Bus Receiver A line receiver intended to be driven from a bus.

Bussing The connecting or joining of two or more circuits.

Butt To join two conductors, end to end, with no overlap and with their axis in line.

Butt Connector A connector in which two conductors come together end to end, with no overlap and with their axis in line.

Butt Contact A mating contact configuration in which the mating surfaces engage end to end without overlap and with their axis in line. The engagement is by spring pressure, with the ends designed to provide optimum surface contact.

Butter Coat A significant thickness of nonreinforced surface-layer resin of the same composition as that within the base laminate material.

Butting Die A crimping die in which the nest and indentor touch at the ends of the crimping cycle.

Butt Joint A connection formed by the end of a lead with the printed wiring board land pattern. *Also called an I-lead.*

Button Hook Contact A contact having a curved, hook shaped configuration. It is often located at the rear of hermetic headers to facilitate soldering and desoldering of leads.

Byte (1) A series of adjacent bits that define a character. (2) A binary character string operated upon as a unit and usually shorter than a computer word. A byte is usually eight bits. *See also Bit.*

C

Cable (1) A stranded conductor with or without insulation or cover. (2) Several conductors insulated from one another as a multiconductor cable. (3) In fiber optics, a jacketed bundle of fibers in a form that can be terminated.

Cable Assembly A cable equipped with connectors or plugs at each end.

Cable Clamp A clamping device attached at the rear of a receptacle, and around the wires, to provide mechanical support.

Cable Terminal A device that seals the end of a cable and provides insulated egress for the conductors. *Sometimes called end bell or pot head.*

Camber The amount of overall warpage in a substrate.

Can Package A cylindrical shaped package with leads attached to one end. *See also TO Package.*

Cantilevered Contact A spring contact in which the contact force is provided by cantilevered springs. It provides a uniform contact pressure and is used in printed wiring board connectors.

Capacitance The property of a system of conductors and dielectrics that permits storage of electricity when potential differences exist among the conductors. The capacitance value is always positive and is expressed as the ratio of quantity of electricity to a potential difference. *Sometimes called capacity.*

Capacitance Density The amount of capacitance available per unit area (pF/mil^2 or mfd/in^2). *Also called sheet capacity.*

Capacitive Coupling The interaction of two or more circuits by means of the capacitance between them.

Capacitor A device whose function is to introduce capacitance into the circuit. It is made of two insulating surfaces and separated by an insulating material or dielectric such as air, mica, glass, plastic film, or oil. It stores electrical energy, blocks the flow of direct current, and allows the flow of alternating current. *Also called condenser.*

Capillary A hollow tube, used as the bonding tool, through which the bonding wire is fed. Pressure from the capillary tool is applied to the wire during the bonding cycle to form the bond. *Also called bonding tool or capillary tool.*

Capillary Tool *See Capillary.*

Captive Device A multipart usually screw-type fastener that retains the loosened components, without separation, when removed from the assembly.

Carbon-Carbon Composite Carbon or graphite fibers that are given structural form by weaving, braiding, or other textile technique and then made dense by adding a carbonaceous matrix. The composites are used for ultra-high-temperature applications. *See also Organic Composites.*

Carbon Fiber Fiber produced by the pyrolysis of an organic precursor fiber, such as rayon, polyacrylonitrile (PAN), or pitch, in an inert atmosphere.

Carbon Tracking A phenomenon wherein a high voltage causes a breakdown on the surface of a dielectric or insulating material and forms a carbonized path.

Card A printed circuit panel that provides the interconnection and power distribution to the electronics on the panel and also provides the interconnection capability to the next-level package. *See also Daughter Card.*

Card Cage A container, equipped with guide rails, that provides compact packaging of printed wiring boards. Card cages are available in various sizes and capable of holding varying numbers of cards. They are equipped with heat sink devices and connectors.

Card Edge Connector *See Edge Board Connector.*

Card Guide A metal or nonmetal guide that provides easier insertion and extraction of a printed wiring board into or from a connector.

Card Insertion Connector *See Edge Board Connector.*

Card on Board A packaging design in which several printed circuit cards are electrically and mechanically connected to printed circuit boards at 90° angles.

Card Rack *See Card Cage.*

Card Slot The lengthwise opening in a printed circuit edge connector that receives the printed circuit board.

Carrier A compartmentalized holder used for storing, transporting, hauling, and testing electronic devices to protect them from physical damage. *See also Waffle Pack.*

Case The bottom portion of a device package, usually a flat pack. It contains one or more cavities and all exit terminals, leads, and pins. The case contributes to the hermetic and environmental protection of the electronic parts within and, at the same time, determines the device form factor.

Cast To embed a component or assembly in a liquid resin, using molds or shells. Curing or polymerization takes place without external pressure. *See also Embed, Pot.*

Castellation (1) A semicircular or crown-shaped metallized surface for making attachments to land patterns, such as those used on ceramic leadless chip carriers. (2) Fluting along the edges of a chip carrier or substrate. *Also referred to as detents, groves, or sockets.* When metallized, castellations provide electrical contacts in an assembly by means of a soldering operation. Their cross section is either semicircular or elongated in a half-slot shape.

Catalyst A chemical that causes or speeds up the cure of a resin, but that does not become a chemical part of the final product. Catalysts are normally added in small quantities. *See also Accelerator, Hardener, Inhibitor.*

Catalytic Curing Curing by an agent that changes the rate of the chemical reaction without entering into the reaction.

Cathode (1) The negative (-) pole in an electroplating bath. Positive (+) charged ions in the plating bath leave the bath or solution and deposit on the metal part being plated (i.e., the cathode). *See also Electroplating.* (2) The electrode to which the forward current flows

within a semiconductor diode or thyristor.

Cathode-Ray Tube (CRT) An electron beam tube in which the beam can be focused to a small cross section on a luminescent screen and controlled to produce a visible pattern.

Cation A positively charged ion that migrates toward the cathode during electrolysis. *See also Ion.*

Caul Plate A rigid plate contained within the vacuum bag to impact a surface texture or configuration to the laminate during cure. *See also Bag Molding.*

Cavity (1) The depression in a mold that usually forms the outer surface of the molded part. Depending on the number of such depressions, molds are designated as single-cavity or multicavity. (2) In microelectronic devices, the cavity is the recessed or hollowed-out area of the package bottom, namely, the site of the device, die, or substrate circuitry when installed.

Cell A single unit capable of serving as a DC voltage source by means of transferring ions during a chemical reaction.

Cellulose A plant carbohydrate used in the synthesis of thermoplastic materials.

Cellulosic Resins A family of resins, such as cellulose acetate, cellulose acetate butyrate, and cellulose acetate propionate. These thermoplastic compounds have good electrical properties.

Centerline Average (CLA) The arithmetic average of measured deviations in a surface profile from a mean centerline located between the peaks and valleys.

Center-to-Center Distance *See Pitch.*

Center-to-Center Spacing *See Pitch.*

Centerwire Break In a wire pull test, the break when the wire fractures at approximately the midpoint between the bonded ends.

Centipoise A unit of viscosity, conveniently and approximately defined as the viscosity of water at room temperature. The following table of approximate viscosities at room temperature may be useful for rough comparisons:

**Viscosity of Materials
at Room Temperature**

Liquid	Viscosity in centipoises
Water	1
Kerosene	10
Motor oil SAE-10	100
Castor oil, glycerin	1000
Corn syrup	10,000
Molasses	100,000

Central Processing Unit (CPU) *See Central Processor.*

Central Processor (CP) The part of the computer system that houses the main storage, arithmetic unit, and special groups. It performs arithmetic operations and controls processing and timing signals. *Also called central processing unit, mainframe, or processor unit.*

Centrifugal Casting A plastic fabrication process in which the catalyzed resin is introduced into a rapidly rotating mold, where it forms a layer on the mold surfaces and hardens.

Centrifuge *See Constant Acceleration.*

Ceramics Inorganic, nonmetallic compounds such as alumina, beryllia, and steatite that are fabricated into parts through heat processing. Their characteristics are produced by subjection to high temperatures. Ceramics are used extensively in microelectronics as parts for components, substrates, or packages.

Ceramic-Based Microcircuits Thick-film screened circuits on a ceramic substrate. The circuits usually contain screened resistors, capacitors, and conductors.

Ceramic Chip Carrier A chip carrier made of alumina, beryllia, steatite, or other ceramic material. *See also Chip Carrier.*

Ceramic Dual In-Line Package (CERDIP) A hollow ceramic dual in-line package consisting of a metal leadframe that is located between two ceramic layers and subsequently sealed by firing with a glass frit. *See also Dual In-Line Package.*

Ceramic Leaded Chip Carrier (CLCC) A ceramic chip carrier with external connections consisting of leads around and down the sides of the package. *See also Chip Carrier.*

Ceramic Matrix Composites A fired ceramic part, which is the matrix, reinforced with certain fibers or flakes to achieve special properties. *See also Metal Matrix Composites, Matrix.*

Ceramic Quad Flat Pack (CERQUAD) A surface-mountable package with straight, outward projecting or formed gull wing leads on four sides.

Cermet A solid homogeneous material made of finely ground metal and dielectric particles in intimate contact. Cermet thin films are normally combinations of dielectric materials and metals.

Cermet Thick Film (CTF) A thick-film deposition formed by firing cermet inks or pastes at high temperatures onto a ceramic or other high-temperature substrate.

CERPACK A ceramic package with leads extending from all four sides, used extensively in surface-mount technology. *Also known as CERQUAD or CERPAK.*

CERQUAD *See CERPACK.*

Certificate of Compliance A document issued by a vendor's quality control department stating that a material or product meets the customer's specifications.

Certification A verification that specified testing has been performed and the requirements have been met.

Channel A thin semiconductor layer, between the source region and drain region, in which the current is controlled by the gate potential. *See also Drain, Gate, Source.*

Channels Paths for providing input to and output from computers or other electronic devices.

Characteristic (1) An inherent and measurable property. Such a property may be electrical, mechanical, thermal, hydraulic, electromagnetic, or nuclear and can be expressed as a value for stated or recognized conditions. (2) A set of related values, usually shown in graphic form.

Characteristic Impedance The impedance that a line offers at any point to an advancing wave of the frequency under consideration.

Charge (1) In electrostatics, the amount of electricity present on any substance that has accumulated electrical energy. (2) The electrical energy stored in a capacitor or battery. (3) The amount of plastic material used to load a mold for one cycle.

Charge Carrier In semiconductors, a mobile conduction electron or mobile hole. *See also Semiconductor.*

Charge-Coupled Device (CCD) A charge-transfer device that stores charge in potential wells and transfers it almost completely as a packet by translating the position of the potential wells.

Chase A viselike tool with adjustable, flexible draw bars for pre-stretching a wire mesh screen prior to installing a thick-film screen frame.

Chassis A metal box or enclosure in which electronic components and printed wiring board assemblies are mounted.

Chelate Compound A compound in which metal is contained as an integral part of a ring structure.

Chemically Deposited Printed Wiring Printed wiring that is formed on a dielectric base material by the reactions of chemicals. *See also Additive Process.*

Chemically Reduced Printed Circuit A printed circuit that is formed by etching conductor patterns on a metal-clad dielectric base material. *See also Subtractive Process.*

Chemical Reversion The tendency of a cured plastic or organic material to soften and return to some stage other than the cured condition. Reversion is caused by humidity, temperature, and pressure, or a combination thereof, and results in degradation of most performance properties of the material.

Chemical Stability The ability of a material to resist change in its molecular structure by a chemical

substance over an extended period of time.

Chemical Vapor Deposition A process whereby circuit elements are deposited on a substrate by chemical reduction of a vapor on contact with the substrate.

Chemorheology The flow and chemical reaction of materials in a polymer.

Chessman In wire bonding, a disk, knob, or lever that is used to control the bonding tool with respect to the substrate.

Chip *See Die.*

Chip and Wire A hybrid technology in which face-up semiconductor chips are bonded and interconnected to substrates by flexible wire bonds.

Chip Architecture The design or structure of an integrated circuit chip achieved by incorporating arithmetic logic units, registers, and other configurations.

Chip Bond *See Die Bond.*

Chip Capacitor A discrete device, rectangular in shape, made of ceramic or tantalum materials that adds capacitance to a circuit and is soldered onto hybrid circuits.

Chip Carrier A low-profile package whose chip cavity or mounting area occupies a major fraction of the package area and whose connections, usually on all four sides,

30

consist of metal pad surfaces (on leadless versions) or leads formed around the sides and under the package or out from the package (on leaded versions). The body of the chip carrier, usually square or of low aspect ratio, is similar to that of a flat pack. *See also Leaded Chip Carrier, Leadless Chip Carrier.*

Chip Circuits Integrated circuits that are interconnected on a substrate to form higher level functions; usually unpackaged and subsequently sealed in a hermetic package.

Chip Component (1) An integrated circuit, diode, transistor, resistor, or capacitor in the form of a chip and used in microelectronic circuits. (2) An unpackaged circuit element, active or passive, for use in hybrid microelectronics.

Chip Design, Depopulated A gate array or cell array chip design that readily lends itself to automatic wire bonding.

Chip Mounting Technology (CMT) Any technology to mount and interconnect bare integrated circuit chips to a substrate without an intermediate packaging step.

Chip on Board (COB) The mounting of chips directly on substrates and subsequent wire bonding, tape-automated bonding, or flip-chip bonding for making electrical interconnections. The chips are then given a glob-top coating.

Chip on Flex (COF) A similar interconnection process to chip on board, except that flexible circuits are used as substrates for the chips and the wire board connections. *See also Chip on Board.*

Chip-Out A cavity-shaped defect in the surface or along the edge of a chip, leaving an exposed active junction. Chip-outs are usually caused by mechanical impact in processing and handling.

Chip Packaging The process of physically locating, connecting, and protecting semiconductor devices in an enclosure.

Chip Resistor A square or rectangular ceramic component of approximately 25 to 100 mil. Chip resistors are made of an inert material with a ruthenium oxide surface.

Chisel A wedge- or chisel-shaped bonding tool used in wedge and ultrasonic bonding of gold and aluminum wires to elements or package leads.

Chlorinated Hydrocarbon An organic compound having hydrogen atoms and, more importantly, chlorine atoms in its chemical structure. Trichloroethylene, methyl chloroform, and methylene chloride are chlorinated hydrocarbons. *Also called halogenated hydrocarbon solvents.*

Chlorofluorocarbons (CFCs) A group of chemicals that are used to clean solder flux and that exhibit serious ozone-depleting properties.

31

See also Ozone-Depleting Chemicals.

Chopped Bonds Wire bonds with a gross deformation that greatly reduces the strength of the bonds.

Chopped Fibers Short fibers, usually fractions of an inch in length, chopped from long, continuous fibers. Commonly used as reinforcement in molded plastics.

Chopped Mat Randomly oriented unwoven fibers cut to various lengths and compacted together by heat and pressure. Fibers are normally several inches long. Commonly used in making glass-reinforced plastic sheets or sheet forms.

Chopped Roving Chopped sections of roving. *See also Roving.*

Chuck In wire bonding, the part of the bonding equipment that holds the unit to be bonded.

Circuit The interconnection of electrical elements and devices to perform a desired electrical function. A circuit must contain one or more active elements (devices) in order to distinguish it from a network.

Circuit Board A sheet of copper-clad laminate material, on which the copper has been etched to form a circuit pattern. The board may have copper on one surface (single-sided) or both surfaces (double-sided). *Also called circuit card, printed circuit board, or printed wiring board. See also Printed Wiring Board.*

Circuit Board Packaging The design and assembly of components on printed circuit boards.

Circuit Element Any basic component of a circuit, excluding interconnections.

Circuit Layout The positioning of the conductors and components prior to photoreduction of the layout to obtain a positive or negative.

Circuit Verifier A test analyzer that electrically stimulates a device and monitors it for shorts and opens, as well as for proper response.

Circular Mil A unit of area equal to the area of a circle whose diameter is 1 mil (0.001 inch). Used in specifying the cross-sectional area of round wires.

Circumferential Crimp A type of crimp in which a force is exerted around the entire circumference of a terminal barrel by the crimping dies and forms symmetrical indentations.

Circumferential Separation (1) A crack or separation in the plating around the circumference of a plated through hole. (2) A crack in the solder fillet around the wire lead or an eyelet, or between a solder fillet and a land.

Cladding (1) The thin metal layer or sheet that has been bonded to the laminate core on one or both of its sides. The result is a metal-clad laminate. (2) In fiber optics, a sheathing of a lower refractive index material around the core of a higher refractive index material, providing optical insulation and protection to the reflective interface.

Clamping Force The force applied to a wire bonding tool to effect a bond.

Clamping Screw A threaded screw or bolt in a terminal block that, when tightened, compresses the wire or conductor against the current bar.

Cleaning Removing contaminants such as fluxes and greases from electronic assemblies. *See also Degreasing, Solvents.*

Clean Room A dedicated manufacturing area in which the air is filtered, dust particle size and quantities are held to various levels, and the temperature and humidity are controlled. The procedure prevents contamination of the unprotected circuits during processing.

Clearance The shortest distance between two lines, objects, tracks, or other points.

Clearance Hole A hole in the conductive pattern larger than, but coaxial with, a hole in the printed circuit base material.

Clearing Time In circuit protection devices, the period of current flow after circuit protector activation. In a fuse, this period is the sum of the melting (or vaporization) time and the arcing time. *See also Melting Time.*

Cleavage Force The imposition of transverse or "opening" force at the edge of an adhesive bond.

Clinch A method of mechanically securing components against a pad area prior to soldering, by bending that portion of the component lead that extends beyond the lip of the mounting hole.

Clinched Lead A wire that is bent to create a spring action against the mounting hole in order to make electrical contact with the conductive pattern prior to soldering.

Clinched-Wire-Through Connection An electrical connection made by a component lead or wire after it is inserted through a hole in a printed circuit board. It is formed or shaped to provide contact with the conductive patterns on both sides of the board and subsequently soldered in place.

Clip Terminal The point where hookup wire is clipped against the connector post.

Clock A device that generates periodic signals from which synchronism may be maintained.

Clocking The arrangement of connector inserts, jack screws, pins, sockets, keys, keyways, and other elements to eliminate mismating or crossmating of connectors. *Also called polarization.*

Closed-Cell Material A cellular material made up of cells that are not interconnected. *See also Open-Cell Material.*

Closed-End Splice The condition in which two or more conductors enter the same end of a barrel terminal.

Closed Entry A design that limits the size of mating parts to specified dimensions, usually in reference to pin and socket contacts.

Closed-Entry Contact A female contact designed to prevent the entry of a pin or probing device having a larger diameter than the mating pin.

Clutter Undesirable and interfering radar returns from objects that are not intended targets.

Coat To cover with a finishing, protecting, or enclosing layer of any compound.

Coated-Metal Core Substrate *See Insulated Metal Substrate.*

Coating A thin layer of material, conductive or dielectric, that is applied over components or the base material for purposes such as moisture absorption prevention, improvement of electrical properties, and protection against chemicals, mechanical abuse, and fingerprints.

Coaxial Cable (1) A cable consisting of a center conductor surrounded by a dielectric and an outer conductor or shield that prevents external radiation from affecting the current flowing in the center conductor. (2) A high-bandwidth cable consisting of two concentric cylindrical conductors with a common axis that is used for high-speed data communication and video signals.

Coaxial Controlled Impedance The condition in which the impedance is essentially constant along the entire length of two or more insulated conductors.

Coefficient of Linear Expansion (CLE) *See Coefficient of Thermal Expansion.*

Coefficient of Thermal Expansion (CTE) The change in dimension (X-Y or Z axis) of a material per unit dimension per 1°C rise in temperature. *See also X-Y Axis, Z Axis.* Average values for some common electronic packaging materials, expressed in 10^{-6} cm/cm/°C, are: See Table 1: CTE of Electronic Packaging Materials on next page.

Cofired Ceramic *See Cofiring.*

Cofiring (1) A process in which thick film screened circuits and resistors are fired simultaneously. (2) A process in which multiple layers of green tape ceramic are fired simultaneously. *See also Green Tape.*

Cohesion (1) The state in which the particles of a single substance are held together by primary or secondary valence forces. (2) In the adhesive field, the state in which the particles of the adhesive (or the adherend) are held together.

**CTE of Electronic
Packaging Materials**

Material	CTE
Silicone elastomers	300
Unfilled epoxy	200
Filled epoxy	75-100
Epoxy-glass laminate	200 (Z axis)
Epoxy-glass laminate	20 (X-Y axis)
Aluminum	20
Copper	20
Glass	5
Kovar	1.2

NOTE: It is this broad range of thermal expansion values that often leads to CTE mismatch and related physical and electrical failures in electronic packages.

Table 1: CTE of Electronic Packaging Materials.

Coil Form Terminals Terminals attached to the coil base or collar to which the coil wires from transformers are connected.

Coined Containing an impression, sometimes unwanted. In thick-film technology, a screen that contains the impression of a substrate is said to be coined. The defect is caused by incorrect printer setup parameters.

Cold Flow The continuing dimensional change that follows initial instantaneous deformation in a nonrigid material under static load at room temperature. *See also Creep.*

Cold Pressing A bonding operation in which an assembly is subjected to pressure without the application of heat.

Cold-Press Molding A molding process in which inexpensive plastic male and female molds are used with room-temperature curing resins to produce accurate parts. Limited runs are possible.

Cold Short A brittle condition in metals that occurs at temperatures below recrystallization.

Cold Solder Connection A soldered joint that was made with insufficient heat or with parts that moved during solidification. It exhibits a grayish, porous appearance and may contain microcracks.

Cold Weld A joining of two metals achieved by pressure alone, such as the forming of a hermetic seal in a metal package by pressure-welding the lid to the frame.

Cold Work The embrittlement of a metal caused by repeated flexing actions.

Collector The element in a transistor that collects the current generated at the junction between the emitter and base. The output part of the transistor.

Collector Electrode A metallized bonding pad in contact with the collector of a transistor element.

Collector Junction The semiconductor junction between the collector and base region of a transistor.

Collimated Light Parallel rays of light, as opposed to converging or diverging rays.

Collocator A mechanical device used to remove substrates from a screen printer and place them in rows on a conveyor or belt for subsequent drying and firing in a furnace.

Colophony A natural product (rosin) whose constituents originate in nature as raw materials. *See also Flux, Rosin.*

Colorant An organic or inorganic dye that is added to a resin for coloring purposes.

Color Coding A system of marking wires, terminals, or contacts with a color to aid in their identification.

Comb Pattern A test pattern in the form of a comb that is applied to a substrate.

Common Part A part that may be used on two or more major items.

Compaction In reinforced plastics and composites, the application of a temporary press bump cycle, vacuum, or tensioned layer to remove trapped air and compact the laminate layers.

Compatible Materials that can be mixed together or brought into contact with one another with little or no reaction or separation.

Compensation Circuit A circuit that alters the functioning of another circuit, with the goal of achieving a desired performance; temperature and frequency compensation are the most common.

Compensation Network *See Compensation Circuit, Network.*

Complementary Integrated Circuit Technology The technology, as applied to integrated circuits, whereby active elements of both polarities are fabricated as monolithic elements on or within the same substrate. For example: A complementary bipolar semiconductor integrated circuit is one that employs both N-P-N and P-N-P bipolar transistors in the same monolithic semiconductor substrate, and a complementary MOS integrated circuit is one that employs both N-channel and P-channel field-effect transistors in the same monolithic semiconductor substrate.

Complementary Metal Oxide Semiconductor (CMOS) (1) A device composed of P-type and N-type MOS used to achieve low power consumption. (2) Logic in which cascaded field-effect transistors (FETs) of opposite polarity are used to minimize power consumption.

Complementary Transistors Two transistors of opposite conductivity (P-N-P and N-P-N) in the same functional unit. They most often have matching electrical characteristics.

Complete Solder Joint (1) A solder joint free of any defects such as nonwetting areas, voids, and improper fillets. (2) A solder connection between a wire and a terminal in which the solder wets and forms a void-free fillet between the wire lead and the complete terminal area.

Compliant Bond A bond that uses an elastically and/or plastically deformable member to impart the required energy to the lead. This member is usually a thin metal foil that is expendable in the process.

Compliant Member The elastically and/or plastically deformable medium that is used to impart the required energy to the lead(s) when forming a compliant bond.

Component An electrical or mechanical part that cannot be disassembled without destroying its intended use. Some typical examples are resistors, transistors, and screws. *See also Discrete Component.*

Component Density The number of components on a substrate or printed wiring board per unit of area.

Component Hole A hole in a printed circuit board used for inserting component leads, pins, and wires.

Component Lead A solid or stranded wire or formed conductor that extends from a component to serve as a mechanical or electrical connector, or both. *See also Stranded Wire.*

Component Side The side of a printed wiring board on which the components are to be mounted or are mounted. *See also Primary Side.*

Composite (1) A homogeneous material created by the synthetic assembly of two or more materials (a selected filler or reinforcing elements and compatible matrix binder) to obtain specific characteristics and properties. (2) Combinations of materials, as opposed to single or homogeneous materials. (3) A material created from a fiber as reinforcement and an appropriate matrix material in order to maximize specific performance properties. The constituents do not dissolve or merge completely but retain their identities as they act in concert. *See also Carbon-Carbon Composites, Ceramic Matrix Composites, Metal Matrix Composites, Organic Composites.*

Compound, Chemical A substance consisting of two or more elements that are chemically bonded by molecular structure and proportioned by weight.

Compression Connector A connector that, after the ends of the conductors have been inserted in a tubelike body connector, is crimped by an external force, resulting in the crimping of the connector and the ends of the conductors.

Compression Molding A technique of thermoset molding in which the molding compound (generally pre-heated) is placed in the heated open mold cavity and the mold is closed under pressure (usually in a hydraulic press), causing the material to flow and completely fill the cavity, with pressure being held until the material has cured. See Fig. 2.

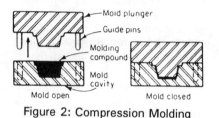

Figure 2: Compression Molding

Compression Seal A seal between an electronic package and its leads. The seal is formed as the heated metal, on cooling, shrinks around the glass seal to create a tight joint.

Compressive Strength The maximum compressive stress that a material is capable of sustaining. For materials that do not fail by a shattering fracture, the value is arbitrary, depending on the distortion allowed.

Computer-Aided Design (CAD) The interactive use of computer systems, programs, and procedures in an engineering design process wherein the decision-making activity rests with the human operator and a computer, which provides the data manipulation functions.

Computer-Aided Engineering (CAE) The interactive use of systems, programs, and procedures in any engineering process wherein the decision-making activity reacts with the human operator and a computer, which provides the data manipulation functions.

Computer-Aided Manufacturing (CAM) The interactive use of computer systems, programs, and procedures in various phases of the manufacturing process wherein the decision-making activity rests with the human operator and a computer, which provides the data manipulation functions.

Computers Electronic machines that can accept, store, and process information mathematically according to instructions, and subsequently provide results after processing the information.

Concentricity In a wire or cable, the measurement of the location of the center of the conductor with respect to the geometric center of the surrounding insulation. *See also Eccentricity.*

Concurrent Engineering (CE) (1) The optimum integration of corporate resources, facilities, organization, and experience into creating successful new products that have high quality at optimum costs and that meet customer requirements. (2) One of the initiatives that is used to reduce the time it takes to achieve optimum product design. (3) A structural process that allows a properly staffed team to create an optimum product design.

Condensation A chemical reaction in which two or more molecules combine with the separation of water, or some other simple substance. If a polymer is formed, the process is called *polycondensation.*

Condensation Polymerization A chemical polymerization in which two or more molecules combine with the separation of other simple substance by-products. *See also Addition Polymerization.*

Condensation Resins Any of the alkyd, phenolaldehyde, and urea-formaldehyde resins.

Condenser *See Capacitor.*

Conductance The ratio of current passing through a material to the potential difference at its ends. The reciprocal of resistance.

Conduction, Thermal The flow of heat in a material from a hot region to a cooler region.

Conductive Adhesive, Electrical An adhesive to which metal particles are added to increase the electrical conductivity. Usually silver particles but sometimes gold or copper are added to the adhesive material.

Conductive Contaminant Growth The electrical bridging or shorting of circuits caused by conductive salts from plating baths, etching solutions, and solder flux residue that arise from inadequate water rinsing or flux removal.

Conductive Epoxy An epoxy resin to which have been added metallic particles such as silver or copper or thermally conductive (dielectric) powders such as alumina and beryllia to increase the electrical or thermal conductivities respectively.

Conductive Filler An electrically or thermally conducting filler material that can be added to plastics and polymers to increase their electrical or thermal conductivity. *See also Filler.*

Conductive Foil A thin sheet of metal that is used to form a conductive pattern on a base material. *See also Foil.*

Conductive Pattern A design formed from an electrically conductive material on an insulating base. Included in this pattern are conductors, lands, and through connections when these connections are an integral part of the manufacturing process, as in the additive process.

Conductive Polymers (1) Polymers that are inherently conductive because of selected doping of the polymer, such as polyacetylene. (2) Polymers that are made conductive by the addition of conductive particles to the polymer, as in conductive epoxies. *See also Conductive Epoxy.*

Conductivity The ability of a material to conduct electrical or thermal energy. Electrical conductivity is the reciprocal of resistivity.

Conductor, Electrical A material that is suitable for carrying electrical current or that has low resistivity ($<10^{-4}$ ohms/cm), such as wire.

Conductor Layer Number 1 The first layer of circuitry that is formed or deposited on the primary side of a circuit board or substrate.

Conductor Side The side of a single-sided board that contains the conductive pattern.

Conductor Spacing The distance between adjacent edges (not centerline to centerline) of printed circuit lines or patterns in a conductive layer of printed circuit.

Conductor Stop A protective device or tool that limits the extension of a wire beyond the conductor barrel.

Conductor Thickness The thickness of a conductor, including all metallic coatings and all protective coatings.

Conductor-to-Hole Spacing The distance between a conductor edge and the edge of a conductor hole.

Conductor Width The width of individual conductors or lines in a conductive film pattern.

Configuration (1) The relative arrangement of parts, such as the components in a circuit. (2) The relative disposition of the external elements of a package, including lead forms.

Confined Crescent Crimp A crimp of two crescent-shaped configura-

tions; one is located at the top and the other at the bottom of the wire barrel crimp.

Conformal Coating A thin dielectric coating applied to the circuitry and components of printed wiring assemblies for environmental protection against moisture as well as fingerprints and for mechanical protection. These coatings can be either a plastic or an inorganic material, and they conform to the surface pattern of the electrical circuitry.

Connection A point in a circuit where two or more components are joined together with nearly zero impedance.

Connection Diagram A drawing showing the electrical connections between the parts and circuits that make up an electronic system.

Connector A device that provides ease in electrical connections and disconnections. Connectors consist of a mating plug and receptacles. Various types of connectors include card edge, two-piece, hermaphroditic, dual in-line package, and wire wrap. Multiple-contact connectors join two or more conductors with others in one mechanical assembly. *See also specific connector types.*

Connector Area The area of a printed wiring board used for providing external electrical connections.

Connector Assembly A mated plug and receptacle.

Connector Block *See Connector Housing.*

Connector Housing A plastic insulating material in which contacts are encapsulated. Once the pins and sockets are inserted in the housing, the assembly is called a *connector*. *Also called a block.*

Connector Insert An electrically insulated molded part that holds the contacts in proper arrangement and alignment. It electrically insulates the contacts from one another and also from the shell.

Connector Insertion Loss The amount of power lost as a result of the insertion of a mated connector onto a cable. *Also called coupling loss.*

Connector Module A group of connector inserts that have the same outside dimensions, can accept different types of contacts, and have different contact shapes.

Connector Set Two or more separate plug and receptacle connectors designed to be mated together.

Connector Shell The casing that holds the connector insert and contact assembly. It provides proper alignment and protection of projecting contacts.

Constant A permanent, fixed, or unvarying value. Usually designated as K.

Constant Acceleration A means of physically testing the integrity of wire bonds, soldered connections, and adhesive bonds in hybrid packages by centrifuging (spinning) at high rates of speed (5000–10,000 RPMs), which imparts a **g** (gravity) loading on bonds and bonded elements. *Also called centrifuge.*

Constraining Core A supporting plane that is internal to a packaging and interconnecting structure.

Constraining Core Substrate A composite printed wiring board composed of epoxy-glass outer layers that are bonded to a low-thermal-expansion core material such as graphite-epoxy, aramid fiber-epoxy, or copper-Invar-copper. The low-thermal-expansion core material constricts the expansion of the outer layer.

Constriction Resistance Resistance due to the contact to a circuit board interface.

Contact The current-carrying member of a connector that engages or disengages to open or close circuits. It provides a separable through connection in a cable-to-cable, cable-to-box, or box-to-box application.

Contact Alignment The amount of excess space or side play that contacts have within the cavity so as to permit self-alignment of mated contacts. *Also called contact float.*

Contact Angle (1) The angle enclosed between a lead and a plane. (2) The angle between a liquid

41

droplet and the solid surface to which it is attached. *See also Low-Surface-Energy Materials.*

Contact Area   The mating surface area between two conductors or a conductor and a connector through which current flows.

Contact Arrangement   The number, spacing, and pattern of contacts in a connector.

Contact Back Wipe   The surface cleaning that occurs during the actuation cycle, as the contacts travel along the mating surfaces and then return on a cleaned, wiped surface at the end of the actuation cycle. *See also Contact Wipe Area.*

Contact Cavity   A defined hole in a connector insert in which the contacts must fit.

Contact Chatter   The vibration of mating contacts resulting in the opening and closing of the contacts.

Contact Durability   A measure of the number of insertions and withdrawal cycles that a connector withstands while meeting its specifications.

Contact Engaging and Separating Force   The force required to engage and/or separate pin and socket contacts when they are not physically located in or out of the connector inserts.   Usually minimum and maximum values are given.

Contact, Female   A contact located in the connection block such that the mating portion is inserted into the unit.   Similar in function to a socket contact. *See also Contact Alignment; Contact, Male.*

Contact Inspection Hole   A hole located in the rear part of a contact used to determine the depth to which a wire has been inserted.

Contact Length   The length of travel of one contact while touching another contact during assembly or disassembly of a connector. *Also called contact mating length.*

Contact, Male   A contact located in the connector block such that the mating portion extends into the female contact.   Similar to a pin contact. *See also Contact, Female; Contact Length.*

Contact Plating   The metal plated on the base material of a contact to provide wear resistance.

Contact Positions   (1) The total number of contacts in a connector. (2) In an edge-type connector, the number of contact positions along the length of the connector.

Contact Pressure   The force that mating surfaces exert on each other.

Contact Printing   In screen printing, the printing that occurs when the screen is almost in contact with the substrate.

Contact Resistance (1) In leaded devices, the apparent resistance between the terminating lead and the body of the device. (2) The resistance between pin and socket contacts of a connector when assembled and in use.

Contact Retainer A device on the contact or in the insert that holds or houses the contact.

Contact Retention Force The minimum amount of force required of a contact to remain engaged within the connector insert. The force is along the contact axis in its normal position within the housing or connector insert.

Contact Shoulder The raised part of the contact that limits its travel distance into the insert.

Contact Size The largest wire gauge size that can be used with the specific contact; it defines the diameter of the engagement end of the pin.

Contact Spacing The centerline distance between adjacent contact areas. *See also Pitch.*

Contact Spring A spring placed inside a socket-type contact during manufacture that exerts a force on the pin to ensure intimate contact.

Contact Wipe Area The contact area over which mating contact surfaces touch during insertion and separation. *See also Contact Back Wipe.*

Contaminant (1) An impurity in a material that can affect its properties. (2) An undesirable particulate that can adversely affect the quality of a product.

Continuity An uninterrupted path for the flow of current in an electrical circuit.

Continuous Belt Furnace A firing furnace equipped with a continuous belt for transporting unfired substrates through a firing cycle.

Continuous Current Rating The stated rms alternating or direct current that a connector or any other electrical device can carry on a continuous basis under specified conditions.

Continuous Use The uninterrupted operation of a component, device, or system for an indefinite period of time.

Control Bus A bus carrying the signals that regulate systems operation within and without the computer. *See also Bus.*

Control Cable A multilead flexible cable intended for carrying signal circuits only where the current requirements are minimum. It is usually relatively small in size and has a relatively small current rating.

Control Chart A means of recording the performance of a process over a period of time in order to identify problems.

Controlled-Collapse Chip Connection
A collapsed solder joint, between a substrate and a flip chip, whose height is controlled by the surface tension of the liquid solder.

Controlled-Collapse Soldering The controlled soldering operation that results in a controlled-collapse chip connection.

Controlled-Impedance Cable A cable containing two or more insulated conductors in which the impedance is nearly constant along its entire length.

Controlled Part A component or device that is made under controlled manufacturing processes and purchased to specified requirements.

Convection A transfer of heat or electricity via moving particles of matter.

Convention A prescribed method used in making electronic diagrams so as to illustrate a clear representation of the circuit function.

Cooling The lowering of the temperature of an object or material.

Coordinatograph A precision drafting machine used to make original artwork for integrated circuits or microcircuits.

Coplanarity (1) The distance between the lowest and highest pin in a package when the package is placed on a flat surface. (2) The

separation between two or more mating surfaces.

Coplanar Leads Flat or ribbon-type leads extending from the sides of an electronic package and all lying in the same plane.

Copolymer A polymer resulting from the chemical reaction of two chemically different substances. The resulting compound has different properties than either of the initial substances. An example is styrene-polyester copolymer.

Cordierite A vitreous ceramic material composed of $2MgO-2Al_2O_3-5SiO_2$ that can be crystallized from glass of the same composition or sintered from powders.

Cordwood Module A high-density package formed by stacking electronic components in cordwood fashion between two sheets of film or other dielectric materials and interconnecting them into electrical circuits by welding or soldering the leads.

Core In fiber optics, the light-conducting, center portion of the fibers bounded by cladding; also, the high-refractive-index region.

Corner Marks *See Registration Marks.*

Corona (1) An electrical discharge, sometimes luminous, stemming from ionization of the air and appearing on the surface of a conductor when the voltage level exceeds a certain value. (2) The flow of small, erratic current pulses resulting from discharges in voids in a dielectric during the voltage stress. *Also called partial discharge*, since corona occurs when the system partially discharges.

Corona Extinction Voltage (CEV) The voltage at which corona discharge ceases as voltage is decreased in a specific set of circumstances. This voltage will not be equal to the corona inception voltage. *See also Corona, Corona Inception Voltage.*

Corona Inception Voltage (CIV) The voltage required to initiate corona as voltage is increased in a specific set of circumstances. *See also Corona, Corona Extinction Voltage.*

Corona Resistance The length of time that a dielectric material withstands the action of a specified level of field-intensified high-voltage ionization without resulting in failure of the insulation. Failure can be erosion of the plastic material, decomposition of the polymer, thermal degradation, or a combination of these three failure mechanisms.

Corrosion A chemical action that causes gradual deterioration of the surface of a metal by oxidation or chemical reaction.

Corrosive Fluxes Fluxes containing inorganic acids and salts that are needed to prepare some surfaces for rapid wetting by the molten solder. *Also called acid fluxes.*

Coulomb The quantity of electricity that passes any point in a circuit in one second when one ampere of current is applied.

Coupled Noise *See Crosstalk.*

Coupler (1) A component used to transfer energy from one circuit to another. (2) A chemical used to improve the bond between a resin and the glass fibers in a composite material. (3) An optical component that interconnects three or more optical conductors.

Coupling Agent A chemical material that can react with both the reinforcement and the resin matrix of a composite or laminate to promote a stronger bond at the interface. The agent may be applied to the reinforcement or added to the resin, or both. *See also Silanes.*

Coupling Capacitor Any capacitor used to couple two circuits together. It blocks DC signals and allows high-frequency signals to pass between parts in an electrical circuit.

Coupling Loss *See Connector Insertion Loss.*

Coupling Ring A device used on cylindrical connectors to lock the plug and receptacle together.

Coupling Torque The force required to rotate a coupling ring when engaging a mating pair of connectors.

Cover (1) A part of a connector that is designed to cover the mating end for mechanical and environmental protection. (2) The top part of an electronic package that is bonded to the bottom part to form a sealed package. A cover has an internal depth, in contrast to a lid, which is flat.

Cover Coat A layer of insulating material, liquid or film, applied over the conductive patterns on the surface of a printed wiring board or flexible circuit. *Also called cover layer. See also Flat Cable, Flexible Printed Wiring.*

Cover Layer *See Cover Coat.*

Cratering A depression or pit under an ultrasonic bond on a chip that was torn loose because of excessive energy transmitted through the wire bond.

Crazing (1) A base laminate condition in which connected white spots or crosses appear on or below the surface of the base material. It is due to the separation of fibers in the glass cloth and connecting weave intersections and is usually related to mechanically induced stress. Similar to measling. (2) Fine cracks that may extend in a network on or under the surface or through a layer of a plastic or glass material.

Creep (1) The dimensional change with time of a material under load, following the initial instantaneous elastic deformation. (2) The time-dependent part of strain resulting from force. Creep at room temperature is sometimes called *cold flow. See also Cold Flow.*

Creepage The conduction of electricity across the surface of a dielectric material.

Creep Distance The shortest distance between two electrical conductors on the surface of a dielectric material.

Crimp To compress or deform a connector barrel around a wire or cable in order to make an electrical connection.

Crimp Contact A contact whose back portion is a hollow cylinder into which a bare wire can be inserted. *See also Solderless Connection.*

Crimper The part of a crimping die that indents or compresses the terminal barrel. *See also Indentor.*

Crimping Chamber The area formed by the mating of the anvil or nest and the crimper or indentor in which a terminal or contact is crimped.

Crimping Die The part of the crimping tool that shapes the crimp.

Crimping Tool A mechanical device used for crimping contacts and terminals.

Crimp Terminal The location where the crimp was made with the bare wire and the pin or contact that mates with the contact terminal.

Crimp Termination A connection in that a metal sleeve is secured to a conductor by mechanically crimping the sleeve with pliers or a crimping machine. *See also Solderless Connection.*

Critical Defect Any anomaly specified as being unacceptable.

Critical Item A part whose failure to meet its designed requirements results in the failure of the product or system. *Also called critical part.*

Critical Part *See Critical Item.*

Cross Connector A connector that joins two branch conductors to the main conductor. The branch conductors are opposite to each other and perpendicular to the main conductor.

Cross Crimp A crimp that applies pressure to the top and bottom of a terminal barrel without collapsing the sides.

Crosshatching The breaking up of large conductive areas by using a pattern of voids to achieve a shielding effect.

Crosslinking The forming of chemical links between reactive atoms in the molecular chain of a polymer. It is crosslinking in the thermosetting resins that makes the resins infusible, strong, and resistant to high temperatures. Thermoplastics, such as polyethylene, can also be crosslinked by irradiation to produce three-dimensional structures that are thermoset in nature and provide increased tensile strength and stress-crack resistance. *See also Thermoset.*

Crossover A point where one conductive path crosses another. The two paths are insulated from each other by a dielectric layer at the area of the crossover.

Crosstalk A type of interference caused by signals from one circuit being coupled into adjacent circuits. Far-end crosstalk is measured by applying and measuring the disturbing signal between two circuits at opposite ends of the cable. Near-end crosstalk is measured by applying and measuring the disturbing signal on two pairs at the same end. *Also called coupled noise.*

Crossunder A crossing of two conductive paths where one path is fabricated into the active substrate for the sole purpose of interconnection.

Crow Foot Weave A type of weave characterized by a 3 x 1 interlacing. That is, a filling thread floats over the three warp threads and then under one. This type of fabric looks different on one side than the other. Fabrics with the crow foot weave are more pliable than those with either the plain or basket weave and, consequently, are easier to form around curves. *See also Basket Weave, Plain Weave.*

Crystalline An atomic or molecular structure that exhibits a regular, patterned, crystallike atomic arrangement. *See also Amorphous.*

Crystalline Melting Point The temperature at which the crystalline structure in a material is breaks down.

Crystalline Polymer A state of molecular structure referring to uniformity and compactness of the molecular chains forming the polymer and resulting from the formation of solid crystals with a definite geometric pattern. In some resins, such as polyethylene, the degree of crystallinity indicates the degree of stiffness, hardness, environmental stress-check resistance, and heat resistance. *See also Amorphous Polymer.*

Crystallization The formation of solids having a definite geometric form and the growth of large crystals from smaller ones. Undesirable or uncontrollable crystallization is called *devitrification.*

C Stage The cured or final stage of a thermosetting resin in which the material will not melt when heated, but may soften. It is insoluble and infusible, and has high molecular weight. *See also A Stage, B Stage.*

C-Stage Epoxy Glass The cured stage of an epoxy resin impregnated glass cloth composite.

Cull Material remaining in a transfer chamber after the mold has been filled. Unless there is a slight excess in the charge, the operator cannot be sure the cavity is filled. The charge is generally regulated to control the thickness of the cull.

Cumulative Distribution Function (CDF) The distribution of a parameter as a part of the total number of measurements with respect to a statistic.

Cure To change the physical properties of a polymer material (usually from a liquid to a solid) by chemical reaction of heat and catalysts, alone or in combination, and with or without pressure. *See also Thermoset.*

Curie Temperature The temperature above which ferromagnetic materials lose their permanent spontaneous magnetization and ferroelectric materials lose their spontaneous polarization. *Also called Curie point.*

Curing Agent Any chemical material that reacts with a base resin, resulting in a final cured or hardened part. *See also Hardener, Thermoset.*

Curing Cycle The total time at a temperature or temperatures required to cure or harden a plastic resin or adhesive to achieve maximum properties.

Curing Temperature The temperature at which a plastic resin or other material is subjected to curing.

Curing Time (1) The period of time during which an assembly is subjected to heat or pressure, or both, to cure the adhesive. (2) The time required for a plastic to completely polymerize or harden.

Curls Extruded bonding material extending from the edge of the bond.

Current The rate of transfer of electricity. It is measured in amperes and represents the transfer of one coulomb per second.

Current Carrying Capacity The maximum current that can be continuously carried by a circuit without causing objectionable degradation of electrical or mechanical properties.

Current Limiting (1) Relating to a type of overcurrent protection device that operates by limiting current to a low and constant value. (2) The property of a fuse that blows (opens) before full short circuit can be delivered to a load by a source.

Current Mode Logic (CML) Integrated circuit logic in which transistors are paralleled to eliminate current drain.

Current Penetration The depth to which current will penetrate into the surface of a conductor at a given frequency. Current flows along the surface of conductors at very high frequencies while deep penetrations exist at low frequencies.

Current Rating The maximum continuous amount of current that a device or component is designed to carry for a specified time at a specified operating temperature.

Custom Design In electronic packaging, a design in which the placement of devices and routing of conductors vary from a standard array but remain within specific tolerances.

Cut and Strip A method of making artwork using a two-ply laminated plastic sheet, by cutting and stripping off the unwanted part of the opaque layer from the translucent layer, thereby leaving the desired artwork configuration.

Cyanate Ester A resin used to manufacture circuit board laminates having higher thermal stability than FR-4 epoxy-glass laminates. *See also BT Laminate, Difunctional Epoxy, Non-Amine-Cured Epoxy.*

Cycle (1) One complete operation of a molding press from closing time to closing time. (2) The change of an alternating wave from zero to a negative peak, to zero, to a positive peak, and back to zero. (3) A regularly repeated sequence of operations.

Cycles per Instruction The number of cycles necessary to process an instruction.

Cycle Time The unit of time that elements of the central processing unit require to complete their functions. One or more cycles may be needed to complete a function. *See also Central Processor.*

Cycloaliphatic Epoxy An epoxy resin that includes a cyclic oxygenated structure as part of the epoxy mole-

49

cule, as compared with a difunctional bisphenol epoxy. This cyclic structure provides a significant

improvement in resistance to high-voltage tracking. *See also Difunctional Epoxy, Tracking.*

D

Daisy Chain A method of device interconnection for determining interrupt priority by connecting the interrupt sources serially.

Damage The failure of an electronic or mechanical component or a material that requires replacement of the defective component or material.

Damping (1) The ability of a material to absorb energy to reduce vibration. (2) The reduction of energy in a mechanical or electrical oscillating system by absorption, conversion into heat, or radiation.

Database A comprehensive collection of information so structured that some or all of its data may be used to create queries about related items contained within it.

Data Bus A bus used to communicate data internally and externally to and from a central processing unit, memory, and peripheral devices. *See also Bus.*

Data Entry Devices A device terminal used to enter information into a computer system.

Data Information A group of records that contain related data describing a specific function or task.

Data Link An information communication channel, normally wideband and digitized.

Datum (1) A theoretically exact point, axis, or plane derived from the true geometric counterpart of a specific datum feature. (2) The origin from which the location or geometric characteristics or features of a part are established.

Daughter Board *See Daughter Card.*

Daughter Card A card that interfaces with a mother board or backplane. *Also called daughter board or daughter substrate. See also Backplane Panel, Mother Board.*

Daughter Substrate *See Daughter Card.*

Dead Face A method used to protect contacts when not engaged. The most common method is to place a cover on the mating ends of connectors to automatically shield the contacts when the connectors are disengaged.

Dead Front The mating surface of a connector designed so that the contacts are recessed below the surface of the connector insulator body to prevent accidental short circuiting of the contacts.

Debugging The elimination of early failures by aging or stabilizing and selectively pretesting the equipment prior to final test.

Decibel (db) (1) A unit of change in sound or audio intensity. (2) In fiber optics, a unit used as a logarithmic measure to describe the attenuation (optical power loss per unit length) in a fiber. (3) The standard unit for expressing transmission gain or loss and relative power levels.

Decorative Laminates High-pressure laminates consisting of a phenolic-paper core and a melamine-paper top sheet with a decorative pattern.

Decoupling Capacitor A capacitance device that filters out transients in a power distribution system.

Defect Any nonconforming, unacceptable characteristic in a unit that requires attention and correction.

Definition (1) The sharpness of a screen printed pattern. (2) The exactness with which the fine-line details of a printed circuit correspond to the master drawing.

Deflashing Pertaining to a wide range of finishing techniques used to remove the flash (excess unwanted material) from a plastic part; examples are filing, sanding, milling, tumbling, and wheel abrading. *See also Flash.*

Deflection Temperature The temperature at which a standard ASTM D-648 plastic test bar deflects 0.01 inch under a load of 66 or 264 psi. *Also called heat deflection temperature (HDT), heat distortion point, or heat distortion temperature.*

Degas To remove air or other gases from a liquid resin mixture, usually by placing the mixture in a vacuum. Entrapped gases or voids in a cured plastic can lead to premature failures, either electrical or mechanical. *Also called evacuate.*

Degradation A gradual deterioration in the performance of a given material or part.

Degreasing The removal of oils and grease from electronic assemblies. *See also Solvents, Vapor Degreasing.*

Deionized Water Water that has been treated to remove ions or ionized material and thereby achieve high resistivity. Deionized water is required for cleaning in certain electronic applications to remove ionic contamination. It has a higher level of purity than demineralized water. *See also Demineralized Water.*

Delamination A separation between any of the layers of the base laminate or between the laminate and the metal cladding originating from or extending to the edges of a hole or edge of the board.

Delay Time (1) The time interval between a reference point on one waveform and a reference point on another waveform. (2) The time interval between a transition at an input and a resultant change at an output. *See also Fall Time, Rise Time.*

Delid The mechanical steps required to remove a lid or cover from a sealed hybrid package.

Demineralized Water Water that has been treated to remove minerals but not necessarily all ions, which are normally found in hard water. *See also Deionized Water.*

Demounted Tape Automated Bonding (DTAB) Tape-automated bonding wherein the inner lead bond is replaced with a bumpless process, and the soldered outer lead bond is replaced with separable pressure contacts. *See also Inner Lead Bond, Outer Lead Bond, Tape-Automated Bonding.*

Dendrite *See Dendritic Growth.*

Dendritic Growth (1) A conductive treelike growth, known as a dendrite, occurring between conductors, usually under the combined influence of electrical energy and humidity. (2) The electrolytic transfer of metal from one conductor to another. The result is a low-resistance path or a short between the bridged conductors. *See also Electromigration, Whisker.*

Denier A numbering system for fibers or filaments that is equal to the weight in grams of a 9000-meter-long fiber or filament. The lower the denier, the finer the yarn.

Density The weight of a material in relationship to its volume, expressed in grams per cubic centimeter, pounds per cubic foot, or other ratio. *See also Specific Gravity.*

Deposited Set down or formed by deposition. Examples are vapor deposition, sputtering, electroless plating, and electrolytic plating. *See also specific terms.*

Deposited Oxide An oxide layer formed on a surface by methods that do not require the substrate to participate in the reaction. Various methods may be employed, including pyrolysis, evaporation, and sputtering. *See also Grown Oxide, Oxidation.*

Deposition A process of applying a coating to a base material by means of vacuum, chemical treatment, screening, or other methods.

Depth of Crimp The distance from the outside diameter of the wire barrel of the connector to the bottom of the indentation made by the indentor. *See also G Dimension.*

Derating Factor (1) A factor used to reduce the current carrying capacity of a wire or cable for application in an environment other than that for which the value was established. (2) A factor used to reduce the theoretical dielectric strength to a practical dielectric strength level for insulating materials in high-voltage applications.

Design (1) To create original sketches, plans, or drawings in order to achieve a specific end. (2) The arrangement of parts or details to produce a complete unit.

Design Allowables Statistically defined (by tests) material properties used for design, usually referring to stress or strain.

Design for Manufacturability (DFM) (1) Design of a product with full consideration of the manufacture and quality of that product during the design cycle. (2) Ensuring that the best design can be efficiently manufactured. *See also Computer-Aided Design, Concurrent Engineering.*

Desmearing *See Resin Desmearing.*

Desoldering A process of disassembling soldered parts. Methods used include wicking, solder sucking, and solder extraction. It is used to repair, replace, or salvage parts.

Dessicant A substance that will remove moisture from materials, usually moisture that has been absorbed onto the surface of a substance. *Also called drying agent.*

Detent (1) A bump or raised projection from the surface of a part. (2) A stop or holding device, such as a pin, that holds a part or controls an assembly in a given position.

Detritus Fragments of loose material produced during laser resistor trimming that remain in the trimmed area.

Device A discrete electrical element, such as a transistor, resistor, or integrated circuit, that cannot be further reduced without destroying or eliminating its function. *See also specific terms.*

Devitrification A process for converting a glassy matter into a crystalline condition.

Devitrify To change from a vitreous to a crystalline condition. *See also Crystalline, Vitreous.*

Dewetting A condition in which liquid solder has not adhered intimately to an area and has pulled back from the base metal, leaving a thin solder film or coating but none of the base metal exposed.

Dew Point The temperature at which water vapor in the air begins to condense.

Diallyl Phthalate (DAP) An ester polymer resulting from the reaction of allyl alcohol and phthalic anhydride. It is a thermosetting plastic that offers outstanding dimensional stability and resistance to humidity and chemicals.

Diamond Pure carbon crystallized in a cubic structure. Special features of importance in electronics are thermal conductivity higher than copper, stabile semiconductor properties, and hardness greater than any other material. *See also Thermal Conductivity.*

Dice The plural of die. *See also Die.*

Dicing The separation of dice from the wafer, usually by sawing. *See also Wafer.*

Die A tiny, uncased integrated device obtained from a semiconductor wafer that can be active or passive, discrete or integrated. Examples are transistors, diodes, and integrated circuits. *Also called chip.*

Die Attach *See Die Bond.*

Die Bond A process in which a semiconductor chip is mechanically attached to a substrate by an epoxy adhesive or a gold-silicon eutectic solder alloy. The die bond is made to the back or inactive side of the chip with the circuit side (face) up. *Also called chip bond or die attach.*

Dielectric *See Insulation.*

Dielectric Absorption The accumulation of electric charges within the body of an imperfect dielectric material when placed in an electric field.

Dielectric Breakdown The complete failure of an insulating material, characterized by a disruptive electri-

cal discharge through the material due to a sudden and large increase in voltage. *Also called voltage breakdown.*

Dielectric Constant (1) The property of a dielectric that determines the electrostatic energy stored per unit volume for unit potential gradient. (2) A material's ability to store a charge when used as a capacitor dielectric. (3) The ratio of the capacitance, using the material in question as the dielectric, to the capacitance resulting when the material is replaced by air. *Also called permittivity or specific inductive capacity.*

Dielectric Isolation Electrical isolation of one or more elements of a monolithic semiconductor integrated circuit by surrounding the elements with an insulating barrier such as semiconductor oxide.

Dielectric Layer A layer of dielectric material or insulation between two conductor layers.

Dielectric Loss The time rate at which electric energy is transformed into heat in a dielectric when it is subjected to a changing electric field.

Dielectric Loss Angle The difference between 90° and the dielectric phase angle.

Dielectric Loss Factor The product of the dielectric constant of a material times the tangent of dielectric loss angle or dissipation factor for that material. *Also called dielectric loss index.*

Dielectric Loss Index *See Dielectric Loss Factor.*

Dielectric Phase Angle The angular difference in phase between the sinusoidal alternating potential difference applied to a dielectric and the component of the resulting alternating current having the same period as the potential difference.

Dielectric Power Factor The cosine of the dielectric phase angle (or sine of the dielectric loss angle).

Dielectric Properties The electrical properties of insulating materials, such as insulation resistance, dielectric strength, dielectric constant, and dissipation factor. *See also specific terms*

Dielectric Sensors Sensors that use electrical techniques to measure the change in loss factor (dissipation) and in capacitance during cure of the resin in a laminate. This is an accurate measure of the degree of resin cure or polymerization. *Also called dielectrometer.*

Dielectric Strength The maximum voltage that an insulating material can withstand before voltage breakdown occurs divided by the distance between electrodes. Usually expressed as a voltage gradient, such as volts/mil. Dielectric strength varies with spacing between electrodes. The ASTM standard specimen thickness (electrode spacing) is 125 mil.

Dielectric-Withstanding Voltage The maximum electrical stress (in volts) that a dielectric material can withstand without failure. *See also Breakdown Voltage.*

Dielectrometer *See Dielectric Sensors.*

Differential Scanning Calorimetry (DSC) A technique that measures the physical transitions of a polymeric material as a function of temperature and compares it with another, similar material that is undergoing the same temperature cycle but not experiencing any transitions or reactions.

Diffused Layer The region of a semiconductor into which impurity dopants have been diffused to a concentration of at least the background concentration. The region is often delineated as a P-N junction. *See also Junction.*

Diffused Resistor The normal semiconductor resistor formed by diffusing a junction-isolated region of proper length, width, and sheet resistance to provide the desired resistance value.

Diffusion (1) A movement of matter at the atomic level from regions of high concentration to regions of low concentration. (2) A thermal process by which controlled minute amounts of dopants are impregnated and distributed into a semiconductor material. *See also Dopant.*

Diffusion Bond *See Solid-Phase Bond.*

Difunctional Epoxy A bisphenol epoxy resin molecule having two epoxide reactant units per epoxy molecule. This is the basic molecular structure of the most commonly used epoxies. One main use in electronic packaging is the manufacture of FR-4 epoxy-glass circuit boards. *See also Multifunctional Epoxy, Trifunctional Epoxy.*

Digital Pertaining to data or information displayed in the form of digits or one of a discrete number of codes. *See also Analog.*

Digital Circuit In switching applications, a type of circuit in which the output assumes one or two states.

Digital Integrated Circuit A type of integrated circuit that is intended to accept particular logic states, changes between logic states, or sequences of logic states at its input terminals and convert these to logic states at its output terminals according to a set of logic equations or function tables. *Also called digital microcircuit. See also Analog Integrated Circuit, Integrated Circuit.*

Digital Microcircuit *See Digital Integrated Circuit.*

Digital-to-Analog Converter A device that converts an input digital signal into a proportional analog output voltage or current. *See also Analog-to-Digital Converter.*

Digitizing The converting of feature locations on a flat plane to digital representation in x,y coordinates.

Diluent An ingredient usually added to a resin formulation to reduce the concentration of the resin or reduce its viscosity. *See also Reactive Diluent.*

Dimensional Change Any change in length, width, or thickness of a solid material.

Dimensional Hole A hole in a printed board whose location is determined by physical dimensions or coordinate values that do not necessarily coincide with the stated grid.

Dimensional Stability (1) The ability of a material to resist changes in size and shape. (2) A measure of dimensional change caused by temperature, humidity, age, stress, and chemical treatment. Expressed as units/unit.

Dimple On semiconductor wafers, a smooth surface depression larger than 3 mil in diameter. *See also Wafer.*

Diode A semiconductor device with an anode and cathode that permits current to flow in one direction and inhibits current flow in the other direction.

Dip Coating The coating of electronic assemblies by dipping them in a liquid coating material. *See also Spray Coating, Vapor Coating.*

Dip-Soldered Terminals All the terminals of a connector that, after being inserted in the holes of a printed wiring board, are soldered in position.

Dip Soldering A process in which component leads that are to be soldered to a printed wiring board are brought in contact with the surface of a static molten solder bath and simultaneously soldered to the conductive paths on the printed wiring board.

Direct-Access Storage Device (DASD) Computer hardware that utilizes magnetic recording on a rotating disk surface. The information subsystem is accessed by means of a movable arm that positions one or more read or write heads along the radius of the disk to the desired track.

Direct Attach The direct attachment of semiconductor chips onto a substrate, as in chip on board (COB) and related direct chip attachment technologies, for interconnecting chips. *Also called direct chip attach. See also Chip on Board (COB).*

Direct Capacitance The capacitance between two conductors as measured through a dielectric layer.

Direct Chip Attach (DCA) *See Direct Attach.*

Direct Contact A contact that is made to a semiconductor chip when the wire is bonded directly over the part to be electrically connected rather than by a lateral path or an expanded contact.

Direct Current (DC) A circuit or voltage source in which the voltage remains constant irrespective of time.

Direct Current Voltage Coefficient A measure of changes in the primary characteristics of a circuit element as a function of the voltage stress applied.

Direct Emulsion In screen printing, a liquid form of emulsion that is applied to a screen, as opposed to a solid film-type emulsion that is transferred from a backing film of plastic.

Direct Emulsion Screen A screen that has been coated with a liquid emulsion material. *See also Indirect Emulsion Screen.*

Direct Metal Mask A metal mask made by chemically etching a pattern into a thin sheet of metal.

Directional Coupler A switch connecting two circuits so that energy moves readily in one direction only.

Disconnect (1) A current-carrying device capable of being separated from its mating part. (2) To open a circuit by separating wires or connections rather than by opening a switch to stop current flow.

Discontinuity A separation or interruption resulting in the permanent or temporary loss or variation of current or voltage.

Discrete Component A component that is fabricated prior to installation, as opposed to those that are screen-printed or vacuum-deposited as part of the film network. Typical examples are resistors, capacitors, diodes, and transistors. *See also Component.*

Discretionary Wiring A technique of interconnecting subarrays on a single wafer in which each subarray is electrically tested by probing and the desired array function is attained by the use of a metallization pattern that connects only usable subarrays.

Dislocations Atomic imperfections or faults in the crystalline lattice structure. The two types are edge dislocations (if a row of atoms is removed or displaced and the slippage is at right angles to the displacement) and screw dislocations (if the slippage is parallel). If the imperfections appear at the surface of the crystal, they are sometimes referred to as *surface dislocations.*

Dispersion (1) The distribution and suspension of very fine particles in another substance. (2) The separating of waves into different frequencies, velocities, and so on.

Displacement Current A current that exists in addition to ordinary conduction current in an AC circuit. It is proportional to the rate of change of the electric field.

Disruptive Discharge The sudden and large increase in current through an insulation medium resulting from complete failure of the medium under electrostatic stress. *See also Breakdown Voltage.*

Dissipation The undesirable loss of electrical energy, generally in the form of heat.

Dissipation Factor A measure of the AC loss, numerically equivalent to the tangent of the dielectric loss angle. *Also called loss tangent and tan delta. See also Dielectric Loss Angle.*

Dissolution Rate The preferred term for the rate at which solid base metal dissolves. *Also called soldering dissolution.*

Distance to Neutral Point (DNP) The distance from the neutral point to the separation of a joint on a chip, with the neutral point being the geometric center of an array of pads on a substrate. It is the point at which there is essentially no motion of the chip and substrate in the xy plane during thermal excursions. This dimension controls the strain on the joint caused by the expansion mismatch between chip and the substrate.

Distributed Element In microwave electronic systems, a distribution of elements or electrical functions to form a circuit. *See also Active Element, Element, Lumped Element.*

Disturbed Solder Joint A joint in which the members to be connected were moved during solidification of the solder.

Doctor Blade (1) A straightedge or knife located above the casting slurry that is used to control the thickness of the slurry. The up-

and-down movement of the doctor blade is controlled by a micrometer, which in turn controls the thickness of the slurry. *See also Slurry, Tape Casting.* (2) A straight piece of material used to spread and control the amount of resin applied to roving, tow, tape, or fabric.

Dome In a cylindrical container, the portion that forms the integral ends of the container.

Dopant A material added to another material by any of various means in order to achieve desired characteristics in the resulting material. Examples are germanium tetrachloride or titanium tetrachloride used to increase the refractive index of glass for use as an optical-fiber core material, and gallium or arsenic added to silicon or germanium to produce a doped semiconductor for achieving donor or acceptor, positive or negative, material for diode and transistor action. *See also Diffusion, Doping.*

Doping The addition of selective impurities to semiconductor materials to alter their conductivity. Common dopants are aluminum, antimony, arsenic, gallium, and indium. *See also Dopant.*

Doppler Effect A change in observed frequency of a wave caused by the time rate of change of the radial component of relative velocity between the source of the wave and the point of observation. An example is the increase in observed frequency of a light or sound wave from a source of constant frequen-

cy increasing its speed toward an observer. If the source continues to accelerate closer, the observed frequency will increase, and if it accelerates away, the frequency will continue to decrease. *See also Doppler Radar.*

Doppler Radar Radar that detects the radial component of the velocity of a distant object (target) relative to the radar antenna by means of the Doppler effect. *See also Doppler Effect.*

Dosimeter A device worn in an environment containing radioactive material for indicating levels of radiation to which people are exposed.

Dot Coding A tool-imprinting process that indicates whether the proper tool has been used.

Double-Grip Terminal A solderless terminal with a metal sleeve that is added to the barrel to provide area for a double crimp. A double crimp is thus made, one over the wire and another over the insulation, to prevent strain from reaching the barrel crimp. These are used in vibration applications.

Double Pole A contact arrangement with two separate contact combinations (i.e., two single-pole contact assemblies).

Double-Sided Substrate A substrate, such as a printed wiring board, that has circuitry on both the top and bottom sides and is interconnected by metallized through holes or edge metallization, or a combination of both.

Drag Soldering A process by which moving, supported, printed wiring assemblies are brought in contact with the surface of a pool of molten solder and the soldered connections are made.

Drain A region of a semiconductor device into which majority carriers flow from the channel. *See also Channel, Source.*

Drain Conductor *See Drain Wire.*

Drain Wire An uninsulated wire in a cable, that is in contact with the shield along its entire length. It is used for terminating the shield and for grounding purposes. *Also called drain conductor.*

Drawbridging *See Tombstoning.*

Dressed Contact A contact with a locking spring permanently attached.

Drift The rate of change in value of an electrical component or device over a period of time as a result of the effects of temperature, aging, humidity, or some other factor. Expressed in percentage per 1000 hours.

Driver An electronic circuit or separate chip that supplies signal voltage and current or input to another circuit.

Dross Oxides and other impurities that form on the surface of molten solder and are periodically skimmed from the surface.

Dry (1) To change the physical state of an adhesive on an adherend through the loss of solvent constituents by evaporation or absorption, or both. (2) To remove moisture from a material or part by heat or evaporation, or both, or in the presence of a dessicant or drying agent. *See also Dessicant.*

Dry Circuit A circuit in which the voltage and current are at a very low level, thereby eliminating any arcing to roughen any contacts. As a result, an insulating film can form and prevent electrical contact when the circuit is closed if the proper means are not employed to prevent formation of the film.

Dry Film Resist A photoresist material in the form of a light-sensitive film, supplied in roll form and processed dry. It is designed for use in the manufacture of printed boards and is resistant to electroplating and etching processes.

Dry Inert Atmosphere An atmosphere in which the water molecules have been removed by the circulation of an inert gas such as nitrogen.

Drying Agent *See Dessicant.*

Dry Pressing The compacting of dry powders with binders in molds under heat and pressure to form a solid mass that is subsequently sintered.

Dry Print Screened resistors and conductors that have been processed through the drying cycle, removing the solvents from the thick-film paste, but not yet fired.

Dry Strength The strength of an adhesive joint determined immediately after drying under specified conditions or after a period of conditioning in the standard laboratory atmosphere. *See also Wet Strength.*

Dual in-Line Package (DIP) An electronic package containing an integrated circuit chip which is connected to terminals or leads that are positioned in two straight rows on the sides of the package, and have standard spacings between the leads and between the rows of leads. These leads are oriented perpendicular to the seating plane for inserting into interconnecting holes in a substrate. See Fig. 3.

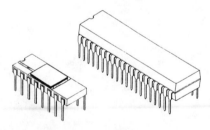

Figure 3: Dual in-Line

Dual-Wave Soldering A soldering process that utilizes both a turbulent wave and a laminar wave. The turbulent wave gets into small, constricted areas to ensure good coverage, while the laminar wave removes solder projections. *See also Wave Soldering.*

Ductility The ability of a material to deform plastically before fracturing.

Dummy Connector Plug A part included in the total design of a connector to mate with a connector but not to perform an electrical function.

Dummy Connector Receptacle A part included in the total design of a connector that mates with a plug connector but cannot be attached to a cable. It is used to seal out moisture.

Dummying A process in which a metal plate having a large total area is placed in an electroplating solution, attached to the cathode, and used for removing impurities from the solution.

Durometer An instrument used to measure the hardness of a nonmetal such as rubber or plastic. Available in different types to cover the ranges of soft, medium, and hard materials.

Duty Cycle The specified operating and nonoperating times of equipment.

Dye Penetrant A liquid material containing fluorescent particles, usually red, and a carrier fluid that is used to detect cracks in solid parts or to locate a gross leak in a sealed electronic package.

Dynamic Burn-In *See Burn-In, Dynamic.*

Dynamic Flex Flexible circuitry used in applications in which continuous flexing is required. In contrast, static flex circuitry remains in a fixed position. *See also Static Flex.*

Dynamic Gap In an edge-type connector, the minimum distance between opposite contacts when the printed wiring board is rapidly removed. The dynamic gap is required to prevent an electrical short from occurring.

Dynamic Mechanical Analysis (DMA) A test method for plastic materials in which dimensional changes are measured with changes in temperature.

Dynamic Printing Force The fluid force that causes a pseudoplastic paste to flow through a screen mesh and wet the surface below.

In thick-film technology, its absolute value is a complex function of all screen-printing operating parameters together with the rheological flow properties of the paste being printed.

Dynamic Random-Access Memory (DRAM) The main electronic memory storage system of large computers, minicomputers, and some microcomputer systems. It periodically requires refresh cycles to restore and maintain information because it employs transient phenomena, typically charge stored in a leaky capacitor.

Dynamic Range The range of useful linear operation expressed as a ratio of the saturation input signal to the noise equivalent signal.

Dynamic Testing Testing of hybrid circuits in which reactions to AC, particularly high frequencies, are evaluated.

E

Early Failure *See Infant Mortality.*

Eccentricity A measure of the center of the location of a conductor with respect to the circular cross section of insulation. It is expressed as a percentage of center displacement of one circle within the other. This term is most commonly used to characterize insulated sound wires. *See also Concentricity.*

Edge Board Connector A one-piece receptacle containing female contacts that mate with printed circuits whose conductors extend to the edge of a printed board. The connector may contain either a single or double row of female contacts. These female contacts are embedded in either a thermoplastic or a thermosetting insulating material. *Also called card edge connector, card insertion connector, or edge connector.*

Edge Board Contacts A series of contacts printed on or near an edge of a printed wiring board and intended for mating with a one-part edge connector. *See also Edge Board Connector.*

Edge Card A printed wiring board having a series of contacts printed on or near any edge or side for the purpose of mating with an edge board connector.

Edge Connector *See Edge Board Connector.*

Edge Definition The reproductive fidelity of the edge of a pattern relative to the production master.

Edge Dip Solderability Test A solderability test in which an edge card specimen is fluxed with a nonactivated rosin flux, immersed in a solder pot at a specific immersion rate for a specific time, and then withdrawn at a specific rate. This test can be by hand operation or a cam-controlled operation. The determination of solderability is made by comparing test results against a standard.

Edge Seal A plug-in package that has a flanged header and cover with a flangeless edge.

Edge Spacing The distance of a pattern or component body from the edge of a printed board.

Egg Crating The insulated walls between each cavity within the contact wire entry face of the connector housing. The wire entry face looks like rectangular cells and minimizes danger of electric shock.

E Glass A family of glasses with low alkali content, usually under 2 percent, most suitable for use in electrical-grade laminates and glasses. Electrical properties remain more stable with these glasses, because of the low alkali content. *Also called electrical-grade glass.*

Eight-Harness Satin A fabric whose weave has a 7 x 1 interlacing in which a filling thread floats over seven warp threads and then under one warp thread. Like the crow foot weave, it looks different on one side than on the other. This weave is more pliable than any of the others and is especially adaptable where it is necessary to form around compound curves, such as on radomes. *See also Basket Weave, Crow Foot Weave, Plain Weave.*

Elastic Deformation A change in the dimensions or shape of a body that is not permanent. *See also Elastic Limit.*

Elasticity The tendency of a material to recover its original size and shape after deformation. If the strain is proportional to the applied stress, the material is said to exhibit *Hookean* or *ideal elasticity*.

Elastic Limit The greatest stress a material is capable of sustaining without any permanent strain remaining when the stress is released. *See also Elastic Deformation, Hooke's Law.*

Elastomer (1) A material that at room temperature can be stretched repeatedly to at least twice its original length and, upon release of the stress, will return with force to its approximate original length.

Plastics with such or similar characteristics are known as *elastomeric plastics* or *thermoplastic elastomers*. A natural or synthetic rubber-reinforced plastic, as in elastomer-modified resins. The resins may be either thermosetting or thermoplastic.

Electrical Relating to any aspect of electricity.

Electrical-Grade Glass *See E Glass.*

Electrical Hold Value The minimum amount of current that will sufficiently energize a relay and maintain electrical contact. *Also called hold current.*

Electrical Isolation The layer of insulation that separates two conductors.

Electrical Losses Losses in electrical performance usually arising from a high dissipation factor or high dielectric constant, or both. Often manifested as generation of heat in the part. *See also Dielectric Constant, Dissipation Factor.*

Electrically Hot Case A hybrid package whose case is part of the grounding circuit.

Electrical Resistance Test A test that measures the resistance in circuits to ensure reliable connections.

Electric Field A region in which a voltage potential exists. The potential level changes with distance and the strength of the field and is expressed in volts per unit distance.

Electric Strength   *See Dielectric Strength.*

Electrode (1) A conductor through which a current enters or leaves an electrolytic cell. This arrangement is commonly found in an electrolytic plating bath or dry cell. *See also Electroplating.* (2) In semiconductors, an element that performs one or more of the functions of emitting or collecting electrons or holes, or of controlling their movements by an electric field. (3) A terminal through which current flows.

Electrodeposition   *See Electrolytic Plating.*

Electroless Deposition   The depositing of a conductive material from an autocatalytic plating solution without an electrical current flowing, usually at room temperature. Electroless deposition can be made on both metallic and nonmetallic base materials. Deposit thickness is limited to below 0.1 mil. *See also Electroless Plating.*

Electroless Plating   The deposition of metallic particles from a chemical solution, usually at elevated temperatures, without an electrical current flowing. This highly controlled process produces uniform thickness but is very time-consuming. Deposits of up to several mils can be produced. *See also Electroless Deposition, Electrolytic Plating.*

Electroluminescence   The direct conversion of electrical energy into light, as in an electroluminescent panel or light-emitting diode.

Electrolytic Corrosion   Corrosion caused by electrochemical erosion.

Electrolytic Plating   A metal deposition process in which an electrolyte — a solution containing dissolved salts of the metal to be plated — transfers cations from the anode into the electrolyte and onto the workpiece or cathode by means of a direct electrical current. *See Electroless Plating.*

Electrolytic Tough-Pitch Copper (ETPC)   A high-purity raw copper possessing the best available physical and electrical properties.

Electromagnet   A coil of wire wound around an iron core that produces a strong magnetic field when the coil is energized.

Electromagnetic   Relating to the combined electrical and magnetic fields that are created by electrons moving through a conductor.

Electromagnetic Compatibility (EMC) (1) The ability of a device or system to function satisfactorily in its electromagnetic environment without producing disturbances to other equipment. (2) The ability of electronic equipment to operate without creating unacceptable electromagnetic interference or responding to such interference beyond specified limits.

Electromagnetic Field A rapidly moving electric field and its associated moving magnetic field, located at right angles both to the electric lines of force and to their direction of motion.

Electromagnetic Interference (EMI) A disturbance in the performance of an electronic system caused by electromagnetic waves that can impair both the reception and transmission of electrical signals. *See also Electromagnetic Shield, Ground Plane.*

Electromagnetic Pulse (EMP) An electromagnetic energy pulse from a nuclear blast or source.

Electromagnetic Radiation (EMR) Radiation resulting from oscillating electric and magnetic fields that propagate with the speed of light. EMR includes gamma radiation, X rays, ultraviolet rays, visible rays, infrared radiation, radio waves, and radar waves.

Electromagnetic Shield A metal screen or metal foil material that surrounds electronic devices or circuits to reduce electric and magnetic fields. The shield is achieved by either reflection or absorption of electromagnetic energy by the material. It is used to prevent electromagnetic interference from entering or leaving a shielded area. *Also called shielding. See also Electromagnetic Interference, Electrostatic Shield.*

Electromigration The migration of electrical energy between conduc-

tors. The result can be a dendritic growth between the conductors. *See also Dendritic Growth.*

Electromotive Force (EMF) The voltage or pressure that causes current to flow in a circuit.

Electron A negatively charge particle that orbits around the nucleus of an atom.

Electron Beam Bonding A process using a stream of electrons to heat and bond two conductors in a vacuum.

Electron Beam Lithography A process in which radiation-sensitive film or photoresist is placed in a vacuum chamber of a scanning-beam electron microscope, exposed by an electron beam under digital computer control, and subsequently developed by conventional processes. Very fine features are possible with this technology.

Electron Beam Welding The process of using a controlled electron beam focused on a small area, converting kinetic energy into high temperatures on impact with the part to be welded.

Electronic Hookup Wire Wire used to make connections between electrical components in electronic assemblies.

Electronic Interference An electrical or electromagnetic disturbance that causes undesirable responses in electronic equipment. *See also Electromagnetic Interference.*

Electronic Package The electro-mechanical assembly resulting from electronic packaging design and manufacture. The level of an electronic package may range from the integrated circuit package assembly to a printed wiring board assembly to a subsystem or system package assembly. *See also Electronic Packaging.*

Electronic Packaging The engineering discipline that combines the engineering and manufacturing technologies required to convert an electrical circuit into a manufactured assembly. These include at least electrical, mechanical, and material design and many functions such as engineering, manufacturing, and quality control.

Electrooptic (EO) *See Optoelectronic.*

Electrooptics That branch of science and technology associated with the interaction of optics and electronics to transform electrical energy into light or light into electrical energy. The result is an optoelectronic device. *See also Optoelectronic.*

Electroplating Plating deposited from a solution by the application of an electrical current. The anode metal is dissolved by chemical and electrical means. Subsequently, cations are deposited onto the cathode (the part to be plated) from the electrolyte. Copper, silver, nickel, chromium, and tin are the metals most commonly electroplated. *Also called electrolytic plating. See also Electroless Plating.*

Electrostatic Charge An electric charge or field on the surface of an insulated object.

Electrostatic Discharge (ESD) The discharge of a static charge on a surface or body of a material or component through a conductive path to ground. An electronic component can be severely damaged if it is not properly grounded to bleed off any static charge before the charge builds up to a high level. *See also Electrostatic Shield, Ground.*

Electrostatic Shield A barrier that prevents the penetration of an electrostatic charge or field. An electrostatic shield offers protection against electrostatic fields, but not against electromagnetic interference. *See also Electromagnetic Shield, Electrostatic Charge.*

Electronic Shielding *See Electromagnetic Shield.*

Electrotinning The electroplating of tin on the surface of a part.

Element (1) The part of a chemical compound that cannot be separated into simpler substances by ordinary chemical means. Such elements contain atoms all of the same atomic number. For example, sodium chloride (NaCl) compound is made up of sodium (Na) and chlorine (Cl) elements. All such elements are displayed, along with their atomic weight, in the periodic table. (2) A topologically distinguishable part of a microcircuit that contributes directly to its electrical characteristics.

Ellipsometer An instrument based on the application and analysis of elliptically polarized light. It is used to measure the thickness of layers and to monitor the quality and composition of thin films, as well as to analyze surface roughness and surface-layer defects. In its automated version, with rapid response time ellipsometry facilitates studies in thin-film growth dynamics.

Elongation The increase in gauge length of a tension specimen, usually expressed as a percentage of the original gauge length. *See also Gauge Length.*

Embed To completely encase a component or assembly in some material, such as a plastic. *See also Cast, Pot.*

Embedded Layer The conductor layer located between insulating layers in a multilayer circuit configuration.

Emissivity The ratio of the radiant energy emitted by a source to the radiant energy of a perfect radiator (black box) having the equivalent surface area, the same temperature, and other identical conditions.

Emitter (1) In relation to transistors, a region from which charge carriers are injected into the base. (2) In optics and electrooptics, a source of electromagnetic radiation in the visible and near visible region of the frequency spectrum. Examples are light-emitting diodes and laser diodes. *Also called source. See also Electromagnetic Radiation.*

Emitter-Coupled Logic (ECL) A form of current mode logic in which the emitters of two transistors are connected to a single current-carrying resistor in a way that only one transistor conducts at a time.

Emitter Electrode The metal pad that is connected to the emitter area of a transistor element.

Emulsion (1) A light-sensitive material used in the photofabrication of printed boards. (2) A light-sensitive material used to coat the screens used in thick-film printing. *See also Photofabrication.*

Emulsion Side As applied to artwork, the side of film or glass on which the photosensitive material is present.

Enameling A process that provides a glassy, virtually pore-free, dielectric finish on metal core substrates or wires. The enamel material may be inorganic, such as porcelain, or organic, such as various polymer films.

Encapsulate To coat or embed a component or assembly with a conformal or thixotropic coating by dipping, brushing, spraying, or potting. *See also Cast, Pot.*

End Bell *See Cable Terminal.*

End of Life (EOL) The wearing out or the end of the useful operating life of a component or system, expressed in a unit of time.

Endothermic Pertaining to a chemical reaction that involves the absorption of heat. *See also Exothermic.*

End Position Mounting A terminal block equipped with end section holes to accept screws without interfering with the contacts.

Ends In wire braiding, the number of essentially parallel wires or threads on a carrier.

End Termination The metallized ends of discrete components, such as capacitors, or the metallized pads on passive chip devices that are used for making electrical contact.

Engaging and Separating Force The force required to engage and separate contact elements in mating connectors. *See also Insertion Force.*

Engineering Change (EC) A change in electrical or mechanical design or a material change.

Entity (1) Two distinct groups of electrical circuits separated by a boundary and by input and output connections. (2) A grouping of parts considered and treated as a single unit.

Entrapment The trapping of air, flux, and particulate in a medium from which they cannot escape.

Environment The combination of temperature, humidity, pressure, radiation, magnetic and electric fields, shock, and vibration that influences the performance of a device, assembly, or system.

Environmental Seal A type of seal to keep out any moisture, air, or dust that might reduce the performance of a device, assembly, or system.

Environmental Stress Cracking The cracking or crazing of a material under the influence of certain chemicals or aging, weather, or stress. Certain thermoplastic materials are especially susceptible to such cracking and crazing.

Environmental Test A series of tests used to determine the external influences affecting the structural, mechanical, and functional integrity of an electronic package, assembly, or system.

Epitaxial Pertaining to a single-crystal layer on a crystalline substrate that has the same crystalline orientation as the substrate.

Epitaxial Growth A process in which layers of materials are grown on a selected substrate. For example, silicon is grown on a silicon substrate or other substrates having compatible crystallographies.

Epitaxial Layer A monocrystalline layer, formed by epitaxy, that is

69

normally of a different resistivity or conductivity type from the substrate material. *See also Epitaxy.*

Epitaxial Peaks Irregular raised points of epitaxial material on the epitaxial surface.

Epitaxy Deposition of a monocrystalline layer of material on a substrate material such that the layer formed has the same crystalline orientation as the substrate. An example is silicon on silicon or silicon on sapphire. *See also Epitaxial Layer.*

Epoxy A thermosetting polymer containing the epoxide group. Most epoxy resins are made by reacting epichlorohydrin with a polyol like bisphenol A. These resins may be either liquid or solid. They exhibit excellent properties for electronic packaging applications.

Epoxy Glass A mixture of glass mat fibers or woven-glass cloth impregnated with an epoxy resin. This is the base structure for the industry standard National Electrical Manufacturers Association (NEMA) Grade FR-4 printed board laminate. *See also FR-4.*

Epoxy Smear Epoxy resin that has been deposited on edges of copper in plated-through-hole circuit boards during drilling, either as a uniform coating or as scattered patches. It is undesirable because it can electrically isolate the conductive layers from the plated-through-hole interconnections. Epoxy smear is commonly removed by chemical etching

or plasma etching. *See also Etchback, Plasma Etching.*

Ester A class of organic compounds formed by the reaction of an acid with an alcohol. Water is a by-product of this reaction.

Etchant A reactive chemical solution used to remove or dissolve away unwanted material.

Etchback The controlled removal of the epoxy smear of the base laminate material by a chemical process acting on the sidewalls of drilled holes. The object is to expose additional internal conductor layers once the epoxy smear is removed. This should not be confused with undercutting. *See also Undercut.*

Etched Metal Mask In screen printing, a thin metal sheet on which a pattern is etched.

Etched Wiring Substrate A printed pattern formed by the chemical removal of conductive material that is bonded to a dielectric base material.

Etch Factor The ratio of the depth of etch to the amount of lateral etch (i.e., the ratio of conductor thickness to the amount of undercut).

Etching A process in which a controlled quantity or thickness of material is removed (often selectively) from a surface by chemical reaction or electrolysis. *See also Plasma Etching.*

Etch Pits Small peaks or holes produced by chemical etching at the site of imperfections in a semiconductor or other surface, caused by the differing etch rate at the point of imperfection.

Ether (1) A nonexisting medium to which wave energy was once considered to propagate. (2) A class of organic materials.

Eutectic A mixture whose melting point is lower than that of any other mixture of the same ingredients.

Eutectic Alloy An alloy having the same temperature for melting and solidus, as shown on the phase diagram.

Eutectic Die Attach The mounting of a semiconductor die to a base material with a preform of a eutectic metal alloy that is brought to its eutectic melting temperature.

Evacuate *See Degas.*

Exotherm The characteristic curve of resin during its cure, which shows heat of reaction temperature versus time. Peak exotherm is the maximum temperature on this curve.

Exothermic Pertaining to a chemical reaction that involves the release of heat. *See also Endothermic.*

Expanded Contact An enlarged contact area formed as an extension to the conductors on the semi-

conductor die for convenience in wire bonding.

Expansion Connector A connector with a built-in flexible connection capability between a rigid conductor and the electrical equipment.

Exponential Failures Failures that occur at an exponentially increasing rate. *See also Wear-Out.*

Exposure The subjecting of photosensitive materials to radiant energy, such as exposing photoresist to ultraviolet light to produce an image.

Extender An inert ingredient added to a resin formulation primarily to increase its volume and/or reduce its cost.

External Leads The flat ribbons or round wires that extend from an electronic package for input or output power, signals, or ground.

External Resistance Thermal resistance from an external point on the outside of an electronic package to a point at ambient temperature.

Extraction Tool A device used to remove contacts from a connector.

Extrinsic Properties Properties in semiconductors that are caused by impurities in a crystal.

Extrinsic Semiconductor A semiconductor with charge-carrier concentration being dependent on impurities. *See also Intrinsic Semiconductor, Semiconductor.*

Extrusion The compacting of a plastic material and the forcing of it through an orifice. The warm, soft extruded shape or form is normally cooled by air or water to harden the plastic. Metals are also extruded in some instances, especially soft metals such as aluminum.

Eyelets Small metal tubes that are used as terminals after being insert

ed in a printed circuit board to provide mechanical support for component leads as well as reliable electrical connections.

Eyelet Tool In ribbon wire bonding, a tool with a square protuberance under the bonding tool surface that presses into the conductor and prevents slippage between the wire or conductor and tool interface.

F

Fabric A planar structure produced by interlacing yarns, fibers, or filaments.

Fabricate To work a material into a finished part by cutting, punching, drilling, tapping, machining, fastening, and other operations.

Fabrication Tolerance The minimum and maximum variations in the characteristics of a part that, when related to other defined variations, will allow operation of the equipment within specified limits.

Face Bonding The process of bonding a semiconductor chip with its circuitry side facing the substrate. Two common face-bonding methods are flip chip and beam lead. The opposite of back bonding. *See also Back Bonding.*

Face-Down Chip A chip intended for mounting with the electrical terminations on the side that is attached to the mounting substrate.

Face Seal A type of design that eliminates any voids at the interface of the plug and receptacle when they are engaged.

Face-Up Chip A chip intended for mounting with the electrical terminations on the side opposite the one attached to the mounting substrate.

Fadeometer Equipment that is capable of accelerating the fading of coatings, resins, and other materials by exposure to high-intensity ultraviolet rays and then determining the resistance of these materials to fading.

Failure A partial or total cessation of the function of an assembly, part, or device as a result of electrical, physical, chemical, or electromagnetic damage. Failure can be temporary or permanent.

Failure Analysis (FA) A determination of the reason for the failure of a part to function at a specified level.

Failure Mechanism A physical or chemical defect that results in intermittent degradation or complete failure.

Failure Mode (1) The manner in which a failure occurs, such as the operating condition. (2) The cause for rejection of any failed device, part, or system as defined in terms of the specific electrical or physical requirement that it failed to meet.

Failure Rate The rate at which devices or parts from a given number of devices or parts can be expected to fail, or have failed as a function of time.

Fall Time The time interval between one reference point on a waveform and a second reference point of smaller magnitude on the same waveform. The first and second reference points are usually 90 percent and 10 percent, respectively, of the steady-state amplitude of the waveform existing before the transition, measured with respect to the steady-state amplitude existing after the transition. *See also Delay Time, Rise Time.*

Fan-In (1) The maximum number of leads that can be connected to the output of a digital device. (2) The ability of a circuit to accept multiple inputs.

Fan-Out (1) The maximum number of leads that can be connected to the input of a digital device. (2) Routing of the electrical path from the chips to the package terminals (e.g., leadframe fan-out for plastic and pressed ceramic packages and screened or plated conductor fan-outs on cofired laminated ceramic and laminated plastic packages). (3) The output ability of a circuit to drive multiple circuits.

Farad A unit of capacitance. When a capacitor is charged with one coulomb, it produces a difference of potential of one volt between its terminals.

Fatigue The weakening of a material caused by the application of stress over a period of time.

Fatigue Factor The factor or force that causes the failure of a material or device when placed under repeated stress.

Fatigue Life The number of cycles of stress that can be sustained prior to failure for a stated test condition.

Fatigue Limit The maximum stress below which a material can presumably endure an infinite number of stress cycles.

Fatigue Strength The maximum stress that can be sustained for a specific number of cycles without failure, with the stress being completely reversed within each cycle unless otherwise stated.

Fatigue Tests Tests that require the application of high stress and a low number of cycles or low stress and a high number of cycles.

Fault Any condition that causes a device or circuit to fail to operate in a proper manner.

Fault Isolation Location of a failure to the level of replacement.

Feedthrough A conductor that extends through the thickness of a substrate or printed board, such as plated through hole or eyelet, thereby electrically connecting circuits on both surfaces. *Also called interface connection.*

Feedthrough Insulator A tubular device made of a nonconductive material that is mounted on a metal chassis or bulkhead and is used to surround a conductor to prevent the conductor from shorting to ground.

Ferrite A powdered, compressed, sintered, magnetic material with high electrical resistivity. It consists of ferric oxide combined with one or more metals.

Ferroelectric Material (1) A nonlinear dielectric material in which electric dipoles line up by mutual interaction. Barium titanate is an example. (2) A material having a high permeability that varies with the magnetizing force. Examples are iron, cobalt, nickel, and their alloys.

Ferrule A short metal tube used to make a solderless electrical connection to a shielded or coaxial cable.

Fiber A relatively short-length, small-diameter threadlike material such as cellulose, wool, silk, or glass yarn.

Fiber Exposure A condition in which glass cloth fibers are exposed on machined, abraded, or poorly laminated areas of a fiber-reinforced material, such as an epoxy-glass laminate having resin-starved areas. *See Resin-Starved Area.*

Fiber Glass An individual filament made from molten glass. A continuous filament is a glass fiber of great or indefinite length; a staple fiber is a glass fiber of relatively short length, generally less than 17 inches.

Fiber-Optic Communication The technology of using light as a means of communication when transmitted through a medium of optical fibers.

Fiber-Optic Connector (1) An assembly used to interconnect a light source to an optical conductor, or an optical conductor to a light detector, or fiber to fiber. (2) A connector that permits coupling and decoupling of optical signals from each optical fiber in a cable to matching fibers in another cable.

Also called optical connector. See also Connector.

Fiber Optics (FO) The sector of optical technology that deals with the transmission of radiant power through glass or plastic fibers. *See also Photon, Photonics.*

Fiber Washout Movement of fiber during cure because of large hydrostatic forces generated in low viscosity resin systems.

Fiducial Marks *See Registration Marks.*

Field Effect Transistor (FET) (1) A semiconductor that is controlled by voltage. The current is controlled between the source terminal and drain terminal by voltage applied to a nonconducting gate terminal. (2) A transistor in which the conduction is due entirely to the flow of majority carriers and in which the conduction can be varied by an electrical field produced by an auxiliary source.

Field-Replaceable Unit (FRU) An electrical subsystem that can easily and readily be replaced in the field or area of operation.

Field Trimming *See Functional Trimming.*

Filament A single fiber of indefinite length.

Filament Winding A process for fabricating a composite structure in which continuous reinforcements, such as filament, wire, yarn, or tape, either previously impregnated with a matrix material or impregnated during the winding, are placed over a rotating and removable form or mandrel in a prescribed way to meet certain product requirements. Generally, the shape is a surface of revolution and may or may not include end closures. When the prescribed number of layers is applied, the wound form is cured and the mandrel is removed. *See also Matrix.*

Fill Yarns that are woven in a crosswise direction in a fabric.

Filled Plastic A plastic material to which have been added ceramic, silica, or metal powders to lower cost or to improve a specific property, such as thermal conductivity, thermal expansion, or strength.

Filler (1) A material used in multiconductor cables to occupy large interstices formed by the assembled conductors. (2) An inert substance added to a plastic to improve properties or decrease cost. *See also Filled Plastic.*

Fillet A concave junction formed at the intersection of two surfaces.

Film (1) A thin sheet of material having a nominal thickness not greater than 0.01 inch. (2) A thin coating or layer of material. *See also Film Casting, Foil, Sheet, Thick Film, Thin Film.*

Film Adhesive A class of adhesives provided in dry-film form with or without reinforcing fabric and cured by heat and pressure.

Film Casting In plastics, the process of making unsupported film or sheet by casting a fluid resin or plastic compound on a temporary carrier, usually an endless belt or circular drum, followed by solidification and removal of the film from the carrier.

Film Conductor An electrically conductive material deposited on a substrate by screen printing, plating, or vacuum deposition. *See also Thick Film, Thin Film.*

Film-Integrated Circuit A nonpreferred term for film microcircuit. *See Film Microcircuit.*

Film Microcircuit A circuit, in film form, made up of elements and formed in place on a substrate by thin- or thick-film techniques.

Film Network An electrical network consisting of thin- or thick-film devices on a substrate.

Film Resistor A film-type device, made of resistive material, on a substrate. *See Thick-Film Resistor.*

Filter (1) A device for removing or separating unwanted electrical signals or functions. (2) A fibrous, organic material used to remove solids and organic impurities from liquid solutions.

Final Seal In microelectronic packaging, the final manufacturing operation that encloses or seals the package so that no further internal processes can be performed without delidding.

Fine Leak A very small leak in a sealed package having a differential air pressure of less than 10^{-7} atm cc/sec of helium. *See also Gross Leak.*

Fine-Leak Test *See Helium Leak Test.*

Fine-Pitch Packages Packages using fine-pitch technology. *See Fine-Pitch Technology.*

Fine-Pitch Technology (FPT) Any interconnection process in which the pitch, or distance between centers of the connections, is between 0.02 and 0.04 inch. *See also Ultra-Fine-Pitch Technology.*

Finger Cots Thin, rubberlike coverings for the fingers, usually made of latex. They prevent contamination of materials being handled in a clean room environment. They can, however, generate static charges when rubbed against an object.

Finger Contact Points on Edge Board Connectors *See Edge Board Connectors.*

Finish A mixture of materials for treating glass or other fibers to reduce damage during processing or to promote adhesion to matrix resins.

Firing A process of thick-film formation whereby the screened film is subjected to a precisely controlled high-temperature condition that oxidizes and vaporizes organic binders and modifies film microstructure to achieve desired properties.

Firing Profile *See Furnace Profile.*

Firing Sensitivity The percentage change in film characteristics during firing as a result of a change in peak firing temperature. It is expressed in units of percentage/degree Celsius. This applies to anything or any type of part.

First Article One of the first parts manufactured that is used for inspection and testing purposes to ensure that requirements are met, prior to manufacturing a larger number of parts.

First Bond The initial bond in a series of two or more bonds to form a conductive connection.

First Radius The radius of the front edge of a bonding tool foot.

First Search The part of the cycle in which any final adjustments are made to the machine in the location of the first bonding area under the tool prior to lowering the tool to make the first bond.

Fish Eye A small area of fabric in a circuit board laminate that resists resin wetting. It can be caused by the resin system, the fabric, or resin impregnation of the fabric. *See also Resin-Starved Area.*

Fish Paper An electrical insulation grade of vulcanized fiber in thin cross section.

Fissuring The cracking of dielectric or conductor materials during firing.

Fixed Contact A contact that is permanently fixed or attached in the insert during the molding process.

Fixed Interconnect Pattern A metallization pattern that interconnects logic elements and is defined by a single, predesigned mask.

Fixed Via A via that exists on a prescribed grid pattern on the various layers of a thick-film multilayer substrate and can interconnect both adjacent and nonadjacent layers.

Flag Terminal A type of terminal in which the tongue extends from the side of the terminal instead of the end of the barrel.

Flame Off The procedure in which a wire is severed by passing a flame across the wire, thereby melting it. Flame-off is used in gold wire thermocompression bonding to form a ball for making a ball bond.

Flame-Retardant Additive A material that when added to a resin mixture, will result in the self-extinguishing of the flame after the source of fire has been removed. *See also Flame-Retardant Resin.*

Flame-Retardant Resin A resin material that is compounded with certain additive chemicals to reduce or

77

eliminate its tendency to burn. *See also Flame-Retardant Additive.*

Flammable Combustible, capable of catching fire, having a low ignition point. *Also called inflammable.*

Flammability The measure of a material's ability to support combustion.

Flange The elliptical part of a connector that extends outside or beyond the periphery of the connector. Its sole purpose is to provide holes to accommodate screws or bolts for mounting to a panel or the mating connector half.

Flanged Spade Tongue Terminal A terminal with a slotted tongue whose ends are bent up or down to prevent the terminal from disengaging from its captive hardware.

Flash During the molding process, unwanted plastic that becomes attached to a molded part along the parting line. The flash must be removed before the part is acceptable for product use. *See also Deflashing.*

Flashover An electric discharge along the surface of a dielectric material, caused by an overvoltage.

Flash Plating A very thin deposition of electroplated material, such as copper or nickel, on a part that will subsequently be plated with a thicker coating.

Flat Base A terminal block equipped with top-to-bottom feedthroughs, but without circuit isolation on the bottom side.

Flat Cable A multiconductor cable assembly whose thin, flat conductors are laid out in the same plane. *Also called flexible cable.*

Flat Cable Connectors Connectors specifically designed to accommodate and terminate flat cable. They are equipped to handle flat conductors, flat cable, or flexible cable with round conductors.

Flat Conductor Cable (FCC) *See Flat Cable.*

Flat Pack (FP) A low-profile package whose leads project parallel to, and are designed primarily to be attached parallel to, the seating plane. The leads typically originate from either two or four sides of a package.

Flex Damage The damage caused by a sharp bend in the wire or cable as it enters the connector housing. To relieve this problem, a flex relief is added to restrict the flexing concentration and cause the wire or cable to bend in a wider arc.

Flexible Cable *See Flat Cable, Flexible Printed Wiring.*

Flexible Conductor Cable (FCC) Rows of parallel conductors that are embedded between flexible plastic sheets or films. This construction forms a cable that can be reproduced in quantity and can be used to replace hand-wired cable bundles, which must be formed individually and are therefore subject to more potential manufacturing errors. *See also Flat Cable.*

Flexible Circuit Carrier A thin, flexible material, such as Kapton, polyamide, or Mylar polyester, on which printed circuits have been deposited by pattern plating. *See also Flexible Printed Wiring.*

Flexible Coating A coating that is still flexible after curing.

Flexible Printed Circuit *See Flexible Printed Wiring.*

Flexible Printed Wiring (FPW) An arrangement of printed wiring utilizing a nonrigid base material or flexible circuit carrier for cover and insulating layers. *Sometimes called flexible printed circuit.* Used to achieve conformability to nonflat structures, such as inside electronic assembly housing or enclosures. *See also Flexible Circuit Carrier.*

Flexibility The quality of a cable or cable component that allows for bending under the influence of an outside force, as opposed to limpness, which is bending as a result of the cable's own weight.

Flexibilizer A material that is added to rigid plastics to make them resilient or flexible. Flexibilizers can be either inert or a reactive part of the chemical reaction. *Sometimes called a plasticizer. See also Plasticizer.*

Flex Life The measurement of the ability of a conductor or cable to withstand repeated bending.

Flexural Strength The strength of a material in bending, expressed as the tensile stress of the outermost fibers of a bent test sample at the instant of failure.

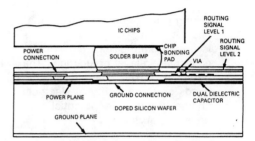

Figure 4: Flip Chip

Flip Chip A semiconductor die having all terminations on one side in the form of solder pads or bump contacts. After the surface of the chip has been passivated, it is flipped over for attachment to a matching substrate. See Fig. 4.

Flip-Chip Mounting A technique of mounting flip chips on thick- or thin-film circuits without subsequent wire bonding.

Flip-Flop A circuit or device that contains active elements capable of assuming either one or two stable states at a given time, depending on input signals and the input terminal receiving the last signal.

Float A warp or fill yarn that does not interlace with the next desig-

79

nated yarn, but passes over or under two or more adjacent yarns.

Floating Bushing An eyelet type bushing located in the plug mounting holes that allows complete freedom of motion in all directions during engagement of the plug and receptacle. Primarily used for easy alignment.

Floating Gate A gate electrode that has no ohmic connection and is isolated from the semiconductor by an insulating layer or junction. The potential on the floating gate depends on the quantity of electrical charge stored in a potential well under the surface.

Floating Ground (1) An electrical ground that is not connected to earth. (2) A ground circuit that does not allow a connection between the power and signal ground for the same signal.

Floating Squeegee In screen printing, a straightedge rubber blade that pushes the ink or paste across the screen and through the mesh onto the substrate and has a rocking ability in the horizontal plane. Some squeegees are rigidly fixed and cannot move during the print and return strokes.

Flood Bar A screen-printing device that drags the paste back to the starting point after the squeegee has printed without pushing the paste through the mesh of the screen.

Flood Stroke In screen printing, the return stroke in which the squeegee redistributes the ink or paste over the pattern on the screen. It is very useful in providing better ink control when thixotropic ink or paste is used.

Floor Planning A layout of the approximate placement and orientation of logic and memory circuitry groupings in a package prior to final design.

Flow Soldering A soldering method that involves the immersion of the metallized areas to be soldered in a molten solder bath to produce an acceptable solder joint.

Fluorinated Ethylene Propylene (FEP) A copolymer of tetrafluoroethylene (TFE) and hexafluropropylene, such as DuPont's Teflon FEP. *See also Tetrafluoroethylene.*

Fluorocarbon An organic compound having fluorine atoms in its chemical structure. This property usually lends chemical and thermal stability to plastics in both liquid and solid form. Typical solid plastic fluorocarbons are tetrafluoroethylene and fluorinated ethylene propylene. *See Fluorinated Ethylene Propylene, Tetrafluoroethylene.*

Fluorochemicals Fluorocarbon liquids having use as heat-transfer materials in electronic cooling systems. *See also Fluorocarbon.*

Flush Conductor A conductor whose longitudinal surface is in the same plane as the adjacent insulating material.

Flush Mount　A type of device whose body is recessed in a panel or chassis and whose face is even with or projects slightly above the surface of the panel or chassis.

Flux　A chemical that attacks and removes oxides from the surface of metals so that the molten solder can wet the surface to be soldered.

Flux Activity　The degree of cleaning and wetting of a metal surface caused by the flux during heating. The flux activity increases continuously during heating, and decreases above a certain temperature because of degradation of the active substances in the flux.

Flux, Foam　A type of flux in the form of foam. In wave soldering, a porous diffuser is used to transform the flux from a liquid to a foam.

Flux, Inactive　A flux that becomes nonconductive after reaching soldering temperatures. *See also Flux.*

Flux, Inorganic Acid (IA)　Flux that is typically composed of hydrochloric, hydrofluoric, or phosphoric acid.

Fluxless Soldering　Soldering under material and process conditions that do not require the use of flux, such as soldering of gold, which does not form oxides, or soldering of adequately cleaned or pretreated metals in nitrogen or other nonoxidizing gaseous environments. *See also Flux.*

Flux, Low-Solids　*See Flux, No-Clean.*

Flux, No-Clean　A type of solder flux that leaves sufficiently low contamination after fluxing that cleaning of the fluxed assembly is not required. This is achieved by the very low solids content of the flux material.

Flux, Organic　*See Flux, Rosin.*

Flux, Organic Acid (OA)　Flux that is typically composed of lactic, oleic, stearic, glutamic, or phthalic acid dissolved in water, organic solvent, petrolatum paste, or polyethylene glycol. *Also called organic water-soluble flux.*

Flux, Organic Water-Soluble　*See Flux, Organic Acid.*

Flux Residue　Particles of flux remaining after soldering and cleaning. If not removed, these residues can lead to various surface failures of the electronic substrate on which they occur. *See also Flux, No-Clean.*

Flux, Rosin (R)　Flux composed of an organic acid, primarily abietic acid, which is a natural organic resin derived from pine tree sap. The rosin is dissolved in isopropyl alcohol, organic solvent, or polyethylene glycol. It is the least active rosin flux and the residue is a hard, transparent film with good electrical insulation properties and resistance to water absorption. *See also Flux, No-Clean.*

81

Flux, Rosin-Activated (RA) A rosin flux that has an even stronger activator than a mildly activated rosin flux.

Flux, Rosin Mildly Activated (RMA) A rosin flux with the addition of a mild activator. Most commonly used in electrical soldering applications.

Flux Solder Connection A solder joint characterized by entrapped flux that often causes high electrical resistance.

Flux, Synthesized-Activity (SA) A flux containing organic derivatives of sulphur and phosphorous acids that are soluble in fluorocarbon solvent blends. The resulting high flux activity is corrosive and requires washing.

Foamed Plastic An insulation material into which an inert gas has been dispersed or which has been made by preblending with a chemical blowing agent to form closed or open cells, thereby reducing the unit weight of the material. *See also Open-Cell Material.*

Foam, In-Place A low-density foam that results from an in-place chemical reaction, usually in a mold. Low-density urethane foams are the most common. *See also Syntactic Foams.*

Foil A thin film of plastic or metallic material, usually under 1 mil (0.001 inch). *See also Film, Sheet.*

Follower A tube or sleeve that compresses the grommet, thereby improving or tightening the seal around the wire as it enters the connector.

Foot Length The long dimension of the bonding surface of a wedge-type bonding tool.

Footprint The area on a substrate intended for the mounting of a particular component and conforming to the geometric pattern of a chip or other component. A group of footprints forms a land pattern. *See also Land Pattern.*

Foreign Material (1) In hybrids, any material that is foreign to the microcircuit or any nonforeign material that is displaced from its original or intended position within the microcircuit package. At preseal inspections, foreign material includes stains from any substance, embedded or attached particles, and loose material. Particles are considered attached when they cannot be removed by brushing lightly or by blowing with a nominal gas flow of about 20 psig. Particulate foreign material in a sealed hybrid device can act as contamination, potentially causing shorts and other degradation. (2) Any unwanted materials or particles in the raw or virgin base material.

Forming Gas Usually nitrogen gas, with small amounts of hydrogen or helium added to the nitrogen, used as a blanket to cover a part and prevent oxidation of metal surfaces.

Forward Bias The bias that tends to produce current in the forward direction. *See also Reverse Bias.*

Forward Current, DC The DC current through a semiconductor diode in the forward direction. *See also Reverse Current.*

Forward Voltage, DC The DC voltage across a semiconductor diode associated with forward current. *See also Reverse Voltage.*

Four-Harness Satin *See Crow Foot Weave.*

FR-4 A fire-retardant epoxy resin glass-cloth laminate, as designated by the National Electrical Manufacturers Association (NEMA). Largely replaces the non-fire-retardant G-10 laminate.

Frame The outermost portion of a connector that has a removable body or insert. It is usually made of metal, supports the insert, and is used for mounting the connector to a panel or the mating connector half.

Frequency The number of complete cycles of some recurring event or phenomenon per unit of time. *See also Hertz.*

Frequency Spectrum The range of frequencies characteristic of the various frequency modes as shown in Table 2.

Fresnel Reflection Losses In fiber optics, the losses that occur at the terminus interface because of differences in refractive indexes.

Fretting A method of maintaining good electrical surfaces of mating contacts by the movement of parts, thereby exposing fresh metal.

Frequency Spectrum and Bands

Frequency designation	Frequency range
•Extremely low frequency (ELF)	30–300 Hz
•Voice frequency (VF)	300–3000 Hz
•Very low frequency (VLF)	3–30 kHz
•Low frequency (LF)	30–300 kHz
•Medium frequency (MF)	300–3000 kHz
•High frequency (HF)	3–30 mHz
•Very high frequency (VHF)	30–300 mHz
•Ultra high frequency (UHF)	300–3000 mHz
•Super high frequency (SHF)	3–30 gHz
•Extremely high frequency (EHF)	30–300 gHz

Radar band frequency

Radar band	Frequency (gHz)
P	0.225 – 0.390
L	0.390 – 1.550
S	1.550 – 3.900
C	3.900 – 6.200
X	6.200 – 10.90
K	10.90 – 36.00
Q	36.00 – 46.00
V	46.00 – 56.00

Table 2: Frequency Spectrum and Bands

Frit Finely ground glass used in thick-film pastes that melts during firing and provides adhesion to the substrates for powdered metals and metal oxides.

From-To List A set of wiring instructions that lists the terminals to be connected.

Front End of Line (FEOL) The first part of the fabrication of devices such as transistors and resistors. The first part includes all the processes used in the manufacture of the integrated-circuit devices.

Front-Mounted Pertaining to a connector or device that is installed or removed from the outside of a panel or chassis.

Front-Release Contacts A type of connector in which the contacts are engaged and released from the front side with the aid of a tool and subsequently pushed out of the back of the connector.

Full Cycling Control A type of crimping tool that cannot be opened until after the crimping has reached its maximum extent.

Fully Additive Process A process wherein the entire thickness of electrically isolated conductors is obtained by the use of electroless and electrolytic deposition. *See also Semiadditive Process, Subtractive Process.*

Functional Test An electrical test in which an assembly is subjected to actual operating conditions.

Functional Trimming The trimming of a resistor to a specific output voltage or current on an operating circuit. *Also called field trimming or trimming.*

Funnel Entry A terminal or connector whose end is flared or funnel-shaped to allow easier insertion of stranded wire into the wire barrel.

Furnace Active Zone The portion of a multizoned muffle furnace that is thermostatically controlled.

Furnace Profile A graph of time versus temperature, or a position in a continuous thick-film furnace versus temperature. *Also called firing profile.*

Furnace Slave Zone A section of a multizoned muffle furnace in which power applied to the heating element is a set percentage of the power supplied to the active zone.

Fuse Terminal Block A terminal block that is equipped with a fuse.

Fusing (1) The melting and cooling of two or more powdered materials so that they bond together in a homogeneous mass. (2) The process of heating and melting a thin metal strip and subsequent resolidification.

G

G-10 A grade of epoxy-impregnated glass cloth printed circuit board material that meets the National Electrical Manufacturers Association (NEMA) requirements. This laminate has been largely replaced with the flame-retardant equivalent NEMA Grade FR-4.

Galvanic Corrosion (1) Corrosion associated with the current of a galvanic cell consisting of two dissimilar conductors in an electrolyte. (2) Accelerated corrosion of a metal stemming from electrical contact with a more noble metal or nonmetallic conductor in a corrosive electrolytic environment.

Galvanic Displacement A chemical deposition of a very thin metallic coating on certain base metals by partial displacement. *Also called immersion plating.*

Gamma Radiation Electromagnetic radiation in the frequency range of $10^{10} - 10^{12}$ gHz. *See also Electromagnetic Radiation.*

Gang Bonding A process in which several mechanical or electrical bonds are made by a single stroke of a bonding tool, as in tape-automated bonding. *See also Tape-Automated Bonding.*

Gang Disconnect A connector that can rapidly connect or disconnect many circuits at one time.

Gas Blanket An inert gaseous atmosphere such as nitrogen flowing over heated parts to prevent oxidation.

Gas-Tight Pertaining to a device capable of resisting the flow of harmful gases or moisture vapor that can cause corrosion.

Gate (1) A circuit with one output and many inputs, designed so that the output is energized only when a specific combination of pulses is present at the inputs. (2) In injection and transfer molding, the orifice through which the melt flows between the runner and mold cavity. *See also Runner, Sprue.* (3) The electrode associated with the region of a semiconductor device in which the electric field created by the central voltage is effective. *See also Channel, Drain, Source.*

Gate Array (1) A geometric configuration of basic gates on a chip. Some arrays contain literally hundreds and thousands of gates, depending on their complexity. (2) A digital integrated circuit containing a configuration of uncommitted elements customized by interconnecting elements with one or more routing layers.

Gauge Length The original of that portion of the specimen over which strain is measured. *See also Elongation.*

Gauntlet A sleeve made of a static-dissipative material with an elastic cuff at one end. It extends from the bare wrist to the elbow and is used to cover long-sleeved apparel not made of a static-dissipative fabric.

Gaussian Pertaining to events that are considered to be in normal distribution, with most events clustered in the center and tapering off at both ends in a pattern known as a bell jar.

G Dimension The distance of the crimped portion of a connector measured between two opposite points on the crimped surface. *Also called T dimension.*

Gel The soft, rubbery mass that is formed as a thermosetting resin goes from a fluid to an infusible solid. This is an intermediate state in a curing reaction, and a stage in which the resin is mechanically very weak. *See also Gelation, Gel Point.*

Gelation The point in resin cure when the resin viscosity has increased to a state where the resin barely moves when probed with a sharp point.

Gel Coat A resin applied to the surface of a mold and gelled prior to lay-up. The gel coat becomes an integral part of the finished laminate, and is usually used to improve surface appearance, and other features.

Gel Point The point at which gelation begins.

Gel Time (1) The time, in seconds, required for a heated prepreg resin to change its physical state from a solid to a liquid and then back to a solid material. (2) The time, in seconds, required for a thermosetting resin to change its physical state from a liquid to a semisolid rubbery material. *See also Gel.*

Geosynchronous Synchronized with the turning of the earth, such as satellites whose period is 24 hours and whose inclination is 0°.

Gerber A file format used by nearly all printed circuit board manufacturers to translate printed circuit board layout files into finished printed circuit boards using photoplotting methods.

Gerber Data A type of data that consists of aperture selection and operation commands and dimensions in x, y coordinates. The data is generally used to direct a photoplotter in generating photoplotted artwork.

Germanium (Ge) A grayish white chemical element with semiconductor properties. Used in transistors and crystal diodes.

General-Purpose Plastics Plastics that cannot be reliably used in operating temperatures above 100°C. *See also Engineering Plastics, High-Performance Plastics.*

Getter A device or material used to capture foreign particles or particulate in a sealed package when the package is vibrated. An example is a gel coating on the underside of the lid or cover of the sealed packaging. In this case, flying particles stick to the gel. *See also Particle Impact Noise Detection.*

Giga- A prefix indicating a multiple of one billion (10^9).

Gigahertz (gHz) Billions of cycles per second. *See also Kilohertz, Megahertz.*

Glass An inorganic, noncrystalline, nonmetallic material made by melting silicon, soda ash, lime, and similar materials.

Glass Binder A glass powder added to resistor and conductor pastes to bind the particles together after firing.

Glass - Ceramic A solid, nonmetallic material with a fine-grain, nonporous microstructure formed by the controlled crystallization of glass.

Glass + Ceramic A solid, nonmetallic material made from mixing crystalline ceramic and glass and subsequently sintering into a composite material.

Glass Fabric A woven cloth made from glass yarns. The glass yarns are made from glass filaments.

Glassivation A protective passivation layer of glass that is applied to surfaces of semiconductor devices. Silicon dioxide is most commonly used as a glassivation layer. *See also Passivation.*

Glass Phase That part of the firing cycle in which the glass binder is in the molten state.

Glass Preform A formed or specifically shaped glass configuration that is either opaque or clear. When pressed into shape with a binder and sintered, the preform forms a hermetic seal in metal packages.

Glass-to-Metal Seal A hermetic seal between glass and metal parts. The seal is made by fusing glass with a metal alloy, such as Kovar, that has nearly the same coefficient of thermal expansion. *See also Kovar.*

Glass Transition Point The temperature at which a material loses its glasslike properties and becomes a semiliquid. *See also Glass Transition Temperature.*

Glass Transition Temperature (Tg) The temperature at which a plastic changes from a rigid state to a softened state. Both mechanical and electrical properties degrade significantly at this point, which is usually a narrow temperature range, rather than a sharp point, as in freezing or boiling.

Glazed Substrate A glass coating on a ceramic substrate that provides a smooth and nonporous surface.

Glazing The application of a glass coating to a ceramic or metal surface, and subsequent firing, which provides a smooth surface and seals the surface against water absorption.

Global Wiring The interconnecting of components that are mounted on a package, not inside the package.

Glob Top The application of a glob of encapsulation material, such as an epoxy or silicone, to the top of a chip after electrical test, wire bond, inspection, or other procedure to completely cover the chip.

Glob-Top Coating A coating process in which a predetermined amount of resin is dispensed on the top surface of a chip or circuit board. After spreading over the desired surface area, the resin coating is cured to form a solid protective coating. This is a common coating process for chip-on-board technology. *See also Chip on Board, Spin Coating.*

Globule Test A test that measures the wettability of a component lead, using a globule of solder, against a known standard of time. *See also Wettability, Wetting.*

Glossy A smooth, shiny, nonporous surface. In thick-film paste materials, it refers to the smooth surface formed by the glass matrix in a conductor or resistor paste.

Glue Line The layer of adhesive that attaches two adherends. *Also*

called bond line. See also Adherend.

Glue Line Thickness The thickness of the fully dried adhesive layer.

Glycol An alcohol containing two hydroxyl (-OH) groups.

Gold Flash A very thin layer of gold, with a thickness of approximately 0.0001 inch or less.

Go/No-Go Test A testing process that yields only a pass or fail condition.

Gradient The rate at which a variable quantity increases or decreases.

Grain Growth The growing in size of the crystal grains in glass or metal over a period of time.

Gram-Force A unit of force required to support a mass of one gram. One gravity unit of acceleration times one gram of mass equals one gram-force.

Graphite Fibers High-strength, high-modulus fibers made by controlled carbonization and graphitization of organic fibers, usually rayon, acrylonitrile, or pitch.

Green Strength (1) The strength of a polymeric substance, joint, or assembly before it has been fully cured. (2) The strength of a ceramic material or part before it has been fired. *See also Green Substrate , Green Tape.*

Green Substrate An unfired, flexible ceramic substrate, that under special conditions is printed prior to firing. Prime examples are cofired multilayer ceramic substrates. *See also Cofiring.*

Green Tape An unfired, flexible ceramic material made from a slurry and having a thin, uniform thickness. This unfired flexibility is due to the presence of soft organic materials in the green tape. Once the flexible green tape is cut or formed as desired, the organic additives are burned out by high-temperature firing. The formed ceramic becomes hard and rigid as a result of the firing process. *See also Alumina Tape.*

Grid (1) A network of equally spaced lines forming squares. (2) An orthogonal network of two sets of parallel equidistant lines that is used for locating points on a printed board.

Grid Array Package A rectangular or square package with terminals attached perpendicular to a major surface in a grid matrix. Prime examples are pad grid array packages and pin grid array packages. *See also Pad Grid Array, Pin Grid Array.*

Grid Spaced The positioning of contacts or pins of a connector in a rectangular configuration such that they are equally spaced from one another.

Grommet A plug-shaped device, usually made of an elastomeric or rubber material, used to seal the rear portion of a connector against moisture, dirt, air, and other harmful contaminants.

Groove The slotted area in a connector that exerts pressure on the cable.

Gross Leak An undesirable leak in a sealed electronic package greater than $10^{-5} cm^3/sec$ at one atmosphere of differential air pressure. *See also Fine Leak.*

Gross-Leak Test A test involving the immersion of a pressurized sealed package in a hot liquid fluorocarbon bath in order to observe rising bubbles, if any, which pinpoint the location of any leaks greater than 10^{-5} atm cc/sec. It is a check on the integrity of a hermetic enclosure. *See also Helium Leak Test.*

Grounded Parts Electrically connected parts that are at the same potential as the earth.

Ground Level The difference in voltage level from the system ground reference point to a second point in the ground system.

Ground Plane A conductive layer on a substrate, printed wiring board, or an inner layer within a substrate that serves as a common reference for electrical circuit return. It also is used for shielding and heat dissipation.

Ground Plane Clearance The etched portion of the conductive material that provides clearance around a plated through hole.

Ground Strap (1) A conductive strap used for dissipating electrostatic charges from conductive pads at work stations to ground. (2) A conductive strap for dissipating electrostatic charges from sensitive electronic parts, in a system or subsystem, to ground. *See also Electrostatic Discharge.*

Grown Oxide An oxide layer formed on a semiconductor surface by the reaction of the semiconductor material with oxygen. *See also Deposited Oxide.*

Guide Pin A hardened or wear-resistant metal pin that extends beyond one mating surface of a two-piece connector to ensure the accurate

alignment of the contacts during the engagement of the connector.

Gull Wing Lead A lead on a surface-mount device that resembles the wings of a seagull. The lead is bent outward, downward, and then upward from the body of the package in order to provide feet for mounting and soldering with inherent mechanical compliance. *See also J Lead.* See Fig. 5.

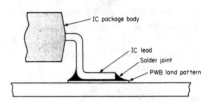

Figure 5: Gull Wing Lead Design

H

Hairpin Mounting A high-density type of mounting in which one lead is bent to form a 180° angle or hairpin configuration, the other lead is left straight, and the axial component is mounted vertically. Not considered reliable in vibration, shock, and impact environments.

Halo (1) An undesirable white ring around a drilled hole in a printed

wiring board. This ring is caused by delamination of the layers, as a result of poor bond strength or by the use of a dull drill. (2) The existence of a glass halo around certain thick-film conductors. It can be avoided by changing furnace profiles or types of materials.

Halogenated Hydrocarbon Solvents Organic solvents containing elements of chlorine, fluorine, bromine, and iodine.

Halogenated Polyester A polyester resin that has been modified with halogens to reduce its flammability.

Halogram An optical image presented on some grated surface in such a way that it appears three dimensional.

Haloing *See Halo.*

Hand Lay-Up The process of placing and working successive plies of reinforcing material or resin-impregnated reinforcement in position on a mold by hand.

Hand Soldering A process in which solder connections are formed by using a hand-held soldering iron for the application of heat.

Hard-Drawn Wire As applied to aluminum and copper, wire that has been cold-drawn to final size so as to approach the maximum strength obtainable and that has not been annealed after drawing.

Hard Wiring Interconnecting wire that is permanently installed and must be unsoldered to disconnect and remove subassemblies.

Hardened Circuit A circuit that can resist damage from transient overvoltages or other forces such as radiation hardening. *See also Radiation Hardened.*

Hardener A chemical added to a thermosetting resin for the purpose

of causing curing or hardening. *Also called curing agent.* Amines and acid anhydrides are hardeners for epoxy resins. Such hardeners are a part of the chemical reaction and a part of the chemical composition of the cured resin. By contrast, catalysts, promoters, and accelerators do not become part of the chemical composition of the cured resin.

Hard Glass A type of glass that has a softening temperature greater than 700°C. The borosilicate glasses are an example.

Hardness A measure of the degree of softness of a material. *See also Barcol Hardness, Indentation Hardness, Rockwell Hardness Number, Shore Hardness Test.*

Hard Solder A solder with a melting point above 425°C.

Hardware Components or devices such as pins, clamps, screws, and terminals. *See also Software.*

Harness A bundle of wires, with or without breakouts, that are held together with ties or covered with a braided or plastic sheath.

Haywire Special insulated magnet wire used primarily for repairs of hybrid circuits, as a jumper to components leads or header pins. *See also Jumper Wire, Magnet Wire.*

Header The base of a package from which the external leads extends. It does not include the lid.

Heat Aging An environmental test in which test articles are heated for predetermined times at predetermined temperatures to indicate the relative resistance to heat degradation. Accelerated aging can often be related to field life, based on proper standards. *See also Accelerated Aging.*

Heat Cleaning The process of removing organic materials from glass cloth by heating to approximately 650 - 700°F for periods up to 50 hours. Heat cleaning can also be used to burn off organic impurities from other thermally stable substrates.

Heat Column The heating element in a eutectic die bonder or a wire bonder. Its purpose is to heat the substrate to bonding temperatures.

Heat Deflection Temperature *See Deflection Temperature.*

Heat Distortion Point *See Deflection Temperature.*

Heat Distortion Temperature *See Deflection Temperature.*

Heat Endurance The length of time and the degree of temperature rise a material or component can withstand before failing a physical or mechanical requirement.

Heat Flux The flow of heat away from a heat source.

Heat Pipe A sealed device used for the transfer of heat by virtue of liquid-to-vapor and vapor-to-liquid heat transfer in the material contained in the sealed device. *See also Thermal Management.* See Fig. 6.

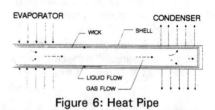

Figure 6: Heat Pipe

Heat Resistance The ability of a material to retain its mechanical and electrical properties as measured during exposure of the material over a temperature range and for a period of time.

Heat Sealing (1) A method of joining plastic films by simultaneous application of heat and pressure to areas in contact. Heat may be supplied conductively or dielectrically. (2) A method for sealing a tape-wrapped wire jacket by thermal fusion of the tape-wrapped layers.

Heat Shrinkable Pertaining to sleeves, tubes, boots, tape, and other forms of plastic that shrink when heated to encapsulate, insulate, and protect connections, splices, and terminations.

Heat Sink (1) Any device that absorbs or transfers heat from the generating source. (2) A separable element or integral part of the package that assists in the dissipation of the heat produced within the package. *Also called heat dissipator.*

92

Heat-Sink Plane A continuous sheet of metal on or in a printed board that provides a path to dissipate the heat away from heat-generating components. *See also Heat Sink.*

Heat Soak The heating of an assembly over a period of time to allow all the circuits, packages, and other components to stabilize in temperature.

Heel The part of the lead adjacent to the bond that is deformed by the edge of the bonding tool in making a wire bond.

Heel Break The fracture or break of the lead at the heel of a wire bond.

Heel Crack A crack across the wire bond width located in the heel area.

Helical Flow Test A test to assess the relative flow of molded plastic materials by measuring, in inches, the length of flow of a plastic sample molded in a helical flow pattern.

Helium A type of gas used in the detection of fine leaks in sealed microelectronic packages. Helium is preferred because of its small molecules and because it is not flammable.

Helium Leak Test A method for detecting fine leaks in hermetically sealed packages in which the trace gas is helium. The leak rate is expressed in terms of cubic centimeters of helium per second at a pressure differential of one atmosphere. *See also Gross-Leak Test.*

Hermaphroditic Connector and Contact Parts that are identical at their mating surfaces.

Hermetic Permanently sealed by fusion or soldering, such as metal to metal or metal to glass, to prevent the transmission of moisture, air, and other gases. Typical maximum leak rates of hermetically sealed packages are 10^{-8}cc of helium per second at a pressure differential of one atmosphere.

Hermeticity The degree of seal of an electronic package to prevent exchange of its internal gas with the external atmosphere. The leak rate of a sealed package is measured in atm cc/sec.

Hertz (Hz) The international system (SI) unit for frequency, equivalent to one cycle per second of some recurring event or phenomenon. *See also Frequency.*

Hex Bar Test A test specified in Military Specification MIL-I-16923 to compare crack resistance of casting and potting resins. *See also Olyphant Washer Test.*

Hidden Via *See Buried Via.*

High-Altitude Electromagnetic Pulse (HEMP) An energy pulse that is caused on or near the surface of the earth by the radiated electromagnetic field from a nuclear blast above the atmosphere.

High Density Pertaining to the number of parts and interconnections in an electronic package. High-density packaging implies fine feature sizes, very close feature spacing, and high input/output (I/O) or pin counts.

High Frequency (HF) In the radio frequency spectrum, the band from 3 to 30 mHz.

High-Frequency Preheating A heating process for plastics wherein the plastic is heated by microwave energy in the frequency range of 20 to 80 mHz. Dielectric loss in the plastic material is the basis for this heating. The process is used for sealing plastic films and preheating molding compounds.

High-K Ceramic A ceramic dielectric composition, such as barium titanate, that exhibits a high dielectric constant and nonlinear voltage and temperature response.

High-Level Noise Tolerance The noise tolerance level of the receiver when the input signal is in its up state.

High-Performance Plastics Plastic materials that are suitable for use above 175°C. *See also Engineering Plastics.*

High-Potential Test *See Hipot.*

High-Power-Microwave (HPM) Devices Microwave devices that exceed 100 megawatts in peak power and span the centimeter and millimeter wave range of frequencies between 1 and 300 gHz.

High-Purity Alumina Alumina having a purity in excess of 99 percent of Al_2O_3.

High-Speed Circuitry Circuitry having a clock rate in the range of 8 to 500 mHz.

High-Surface-Energy Materials Materials that have surface-free energies in the range of 5000 to 500 ergs/cm^2 and are easily wetted or bonded. Typical examples are metals, metal oxides, nitrides, and glasses. *See also Low-Surface-Energy Materials.*

High Temperature Reverse Bias Test An accelerated screening test conducted on microelectronic devices for early detection and elimination of defects that could occur in later system use, causing poor reliability and premature system failure. This test uses predetermined cyclic temperatures and voltage bias levels that are higher than system use, with test times being much less than expected system life. For example, a hybrid microelectronics assembly tested for 168 hours at 125°C without failure could be expected to have a system life of over one year, or nearly 10,000 hours of useful life. Types of failures that can be detected by this test include intermittent shorts and opens, poor wire bonds, near opens, semiconductor circuits with crystal dislocations, diffusion defects, contamination-based problems, improper doping levels of the semiconductor device, cracked dies or chips, and any failures caused by ion migration and dissimilar metal contacts. *See also Bias Voltage; Burn-In; Burn-In, Dynamic; Burn-In, Static.*

High Temperature Stability The ability of a material to perform reliably at high temperatures without major changes in its properties caused solely by heat.

High Voltage (1) An alternating current (AC) circuit whose difference in potential is at least 1 kV rms. (2) A direct current (DC) circuit that functions at a difference of at least 1.4 kV.

Hi-K High dielectric constant.

Hipot An electrical test that measures the voltage breakdown of a dielectric material. *Also called high-potential test.*

Hi-Rel High reliability. Pertaining to a device or component designed and manufactured to close tolerances, by controlled processes, and inspected by rigid requirements to provide a long service life.

Hold Current *See Electrical Hold Value.*

Holding Strength The ability of a connector to remain assembled to a cable or to the individual wires when under tension.

Hole The absence of an electron in the covalent bond of an atom within a semiconductor crystal. The absent electron acts as a positive charge. Consequently, hole conduction in semiconductors is equiv-

alent to the motion of positive charges.

Hole Breakout An unwanted condition in which a hole is only partially surrounded by a land.

Hole Density The number of holes in a unit area of a printed board.

Hole Diameter (1) The diameter of a hole in a printed board, including the metal plating. (2) The diameter of the hole through the bonding tool.

Hole Location The dimensional position of the center of a hole on a printed board.

Hole Pattern The arrangement of all holes in a printed board with respect to a reference point.

Hole Pull Strength The load or pull force along the axis of a plated through hole that will rupture the hole.

Hole Void A void in a plated though hole or metallization exposing the base material.

Homogeneous Pertaining to a material whose composition is uniform throughout.

Homopolymer A polymer resulting from polymerization of a single monomer. *See also Monomer.*

Honeycomb A manufactured product of resin-impregnated sheet material or metal foil formed into hexagonal cells. Skins are bonded to the top and bottom surfaces of the cell structure to achieve strength.

Hood An enclosure on the rear of a connector that contains and protects wires and cable attached to the terminals. A cable clamp is considered part of the hood.

Hooke's Law The physics principle stating that the strain produced in an elastic body is proportional to the stress which causes such strain.

Hook Terminal A terminal with a hook-shaped tongue. *See also Hook Tongue.*

Hook Tongue A terminal with a tongue that opens from the side rather than from the end.

Hookup Wire Insulated wire used for low-current and low-voltage levels (less than 100 volts), for applications in electronic equipment.

Horn A cone-shaped member that transmits ultrasonic energy from the transducer to the bonding tool of an ultrasonic bonder. *See also Ultrasonic Bonding.*

Hostile Environment An environment that degrades electronic circuits, devices, and systems. Examples of hostile environments are high and low temperatures, high humidity, radiation, and vibration.

Hostile Environment Connectors Connectors that are designed and engineered for operation in temper-

atures ranging from absolute zero to 675°C and under high-humidity conditions.

Hot-Air Leveling The smoothing or leveling of a solder surface by a blanket of forced hot air.

Hot-Bar Soldering A process in which a heated bar solders all the leads of a device to the pads of a printed wiring board or substrate simultaneously.

Hot-Electron Diode *See Schottky Diode.*

Hot-Gas Reflow A process in which a heated gas is used to reflow solder and form solder joints and interconnections. *See also Reflow Soldering.*

Hot-Line Connector A connector that may be installed or disconnected by using an insulated stick or other device while the conductor is energized. *Also called a live-line connector.*

Hot-Melt Adhesive A thermoplastic adhesive compound, usually solid at room temperature, that is heated to a fluid state for application.

Hot Spot A point of maximum temperature on a component that is unable to dissipate heat into the surrounding areas.

Hot-Tin Dipping A technique in which a coating of solder is applied to metal parts, such as component leads and printed wiring boards, by immersing the parts in a molten solder bath.

Hot Zone The part of a continuous furnace or kiln that is at maximum temperature and located between the preheat and cooling zones.

Housing (1) The shell portion of a connector minus the insert, but including the positioning hardware. (2) The encasement that contains electronic assemblies.

Human Engineering The science that deals with the designing, building, and testing of mechanical devices to meet the anthropometric, physiological, or psychological requirements of the people who use them.

Humidity The amount or degree of moisture in the air.

Humidity Resistant Pertaining to the ability of a material to resist the absorption of humidity or moisture.

Hybrid A mixture of two or more different technologies and components. Examples are thin- and thick-film circuits and active and passive devices.

Hybrid Circuit *See Hybrid Integrated Circuit.*

Hybrid Device *See Hybrid Integrated Circuit.*

Hybrid Electronic Assembly An electronic assembly containing discrete integrated circuit devices and other chip-type components, thin- or thick-film interconnecting circuits, and wire bonding or other techniques for component attachment and interconnection.

Hybrid Electronics A technology utilizing thin- and thick-film circuitry, discrete integrated circuit devices, and other chip-type components, with wire bonding or other techniques for interconnecting purposes to form electronic circuits. *See also Hybrid Electronic Assembly.*

Hybrid Group Hybrid microcircuits designed to perform the same types of basic circuit functions for the same supply, bias, and voltages.

Hybrid Integrated Circuit A microcircuit made up of thin-film or thick-film circuits on a substrate to which active and passive microdevices are mounted and bonded. *Also called hybrid circuit, hybrid device, or hybrid microcircuit.*

Hybrid Microcircuit *See Hybrid Integrated Circuit.*

Hybrid Microelectronics The electronic art that applies to electronic systems using hybrid circuit technology.

Hybrid Microwave Circuit *See Microwave Integrated Circuit.*

Hybrid Module A special enclosure containing hermetically sealed hybrid packages, discrete passive components, and/or conformably coated nonhermetic hybrid assemblies, all of which are electrically interconnected.

Hybrid Package *See Multichip Package.* See Fig. 7.

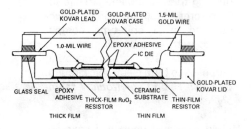

Figure 7: Hybrid Package

Hydrocarbon An organic compound having hydrogen and carbon atoms in its chemical structure. Most organic compounds are hydrocarbons. Aliphatic hydrocarbons are straight-chained hydrocarbons, and aromatic hydrocarbons are ringed structures based on the benzene ring. Methyl alcohol and trichloroethylene are aliphatic; while benzene and toluene are aromatic.

Hydrolysis Chemical decomposition of a substance through the addition of water.

Hydrolytic Stability The degree of resistance of a polymer to perma

nent property changes from combined moisture and temperature effects.

Hydrophilic Having an affinity to absorb water or be wetted by water.

Hydrophobic Having no affinity for water, nonwettable, water repellent. Examples of hydrophobic materials are Teflon and waxes.

Hydroxyl Group A chemical group consisting of one hydrogen atom plus one oxygen atom. The chemical symbol is OH.

Hygroscopic Having a tendency to absorb and retain moisture from the atmosphere.

Hysteresis An effect in which the magnitude of a resulting quantity is different during increases in the magnitude of the cause than during decreases arising from internal friction in a substance and accompanied by the production of heat within the substance. Electric hysteresis occurs when a dielectric material is subjected to a varying electric field, as in a capacitor in an alternating current (AC) circuit.

I

Icicle A nonpreferred term for a solder projection. *See Solder Projection.*

Identification Pin A method of denoting pin 1 in a plug-in package. Usually done by coloring the glass bead or by placing a die-stamped number on the header or on a square corner.

Image Processing The treatment or manipulation of an image by some means. *See also Imaging.*

Imaging The sensing of objects or patterns and their representation by optical, electronic, or other means on a film, screen, platter, cathode-ray tube, electroluminescent display panel, or other device.

Immersion Plating The process used to chemically deposit a thin metallic coating over certain basic metals by a partial displacement of the basic metals. *See also Electroless Plating, Galvanic Displacement.*

Immiscible Pertaining to fluids that will not form a homogeneous mixture or that are mutually insoluble.

Immunity The ability of a system to resist disturbances.

Impact Resistance The resistance of a material to fracture under shock or impact.

Impact Strength The strength of a material, measured in foot-pounds, when subjected to impact forces or loads. *See also Izod Impact Test.*

Impedance The total opposition that a circuit offers to the flow of alternating current or any other varying current at a particular frequency. It is a combination of resistance R and reactance X, measured in ohms and designated by Z.

$$Z = (R^2 + X^2)^{1/2}$$

Impedance Match The condition that exists when the impedance of a component, circuit, or load is equivalent to the internal impedance of the source, or to the surge impedance of a transmission line.

Impregnate To force resin into every interstice of a part. Cloths are impregnated for laminating, and tightly wound coils are impregnated with liquid resin using air pressure or vacuum as the impregnating force.

Impulse A power surge of unidirectional polarity for a given time. *See also Impulse Ratio, Impulse Strength, Impulse-Withstanding Voltage.*

Impulse Ratio The ratio of the flashover, sparkover, or breakdown voltage of an impulse to the peak value of the power-frequency, sparkover, or breakdown voltage. *See also Impulse, Impulse Strength, Impulse-Withstanding Voltage.*

Impulse Strength The area under the impulse amplitude-time curve. *See also Impulse Ratio, Impulse-Withstanding Voltage, Pulse.*

Impulse Withstanding Voltage The peak value of an applied impulse voltage that, under specific conditions, does not cause flashover, puncture, or disruptive discharge on the part under test.

Impurity Diffusion A process used to introduce desired impurities into a semiconductor crystal to alter its electrical properties. Diffusion is accomplished at high temperatures by introducing suitable dopants to the surface of the semiconductor wafer under precisely controlled conditions.

In-Circuit Test An electrical test that is performed on individual components even though they are soldered in place.

Inclined-Plane Furnace A furnace whose hearth is inclined such that a draft of oxidizing atmosphere will flow through the heated zones by natural convection.

Inclusion A foreign particle, metallic or nonmetallic, in a conductive layer, plating, or the base material.

Incomplete Bond A bond impression that is less than normal size because part of the bond is missing.

Indentation Hardness Hardness evaluated from measurements of area or indentation depth caused by pressing a specified tool or indentor into the material surface with a specified force. *See also Barcol Hardness, Shore Hardness.*

Indentor (1) A part of a crimping die set. It shapes the terminal barrel to the desired configuration while the nest provides the location and support of the crimping process. (2) The part of an indentation hardness test that penetrates the surface of the material being tested. *See also Indentation Hardness.*

Index-Matching Materials In fiber optics, those materials that are placed between the ends of the optical conductors to reduce the coupling losses.

Index of Refraction The ratio of the velocity of a light wave in a vacuum to that in a specified medium.

Indirect Emulsion An emulsion that is transferred to the screen surface from a plastic carrier or backing material. *See also Direct Emulsion.*

Indirect Emulsion Screen A screen whose emulsion is a separate sheet or film of material, attached by pressing into the mesh of the screen. *See also Direct Emulsion Screen.*

Induced Electrostatic Discharge An electrostatic discharge (ESD) event brought about on an ESD-sensitive (ESDS) device through polarization and charge accumulation on its surface and induced by the force lines of an electrostatic field emanating from a charge body or an approaching person at a distance. The induced ESD will occur as a subsequent, unintended discharge. *See also Electrostatic Discharge.*

Inductance (L) The property of a circuit or circuit element to oppose a change in current flow, causing current changes to lag behind voltage changes. Measured in henrys.

Induction Soldering A method of soldering in which the solder is reflowed when the work is moved slowly through an electromagnetic field.

Inductive Components Components that create an inductive field when voltage is applied, although that may not be their primary function.

Inductive Coupling The interaction of two or more circuits by means of either mutual inductance or self-inductance common to the circuits.

Inductive Crosstalk Crosstalk created by the coupling of the electromagnetic field of one conductor on another. *See also Crosstalk.*

Inert Atmosphere A gas atmosphere such as that of helium or nitrogen, or a mixture of gases, that is nonoxidizing or nonreducing in the treatment of metals.

Infant Mortality The time regime during which hundreds of circuits may be failing at a decreasing rate, usually during the first few hundred hours of operation. These failures are often detected by an accelerated test such as burn-in or high temperature reverse bias. *Also called early failure. See also Burn-In, High Temperature Reverse Bias Test.*

Inflammable *See Flammable.*

Infrared Radiation (IR) Electromagnetic radiation in the range of frequencies from the longest wavelength of visible red to the radio wave region. This infrared (IR) band lies between the longest wavelengths of the visible spectrum (about 0.8 microns) and the shortest radio waves (100 microns). The IR region of the spectrum is divided into near IR (0.8 – 3.0 microns), middle IR (3 – 30 microns), and far IR (30 – 100 microns). *See also Electromagnetic Radiation.*

Infrared Reflow A process in which an infrared light source is used to heat solder to its melting point in an infrared furnace. Printed wiring boards with surface-mount devices and other types of packages and components, properly positioned, are often soldered by this process.

Inhibitor A chemical added to resins to slow down the curing reaction. Inhibitors are normally added to prolong the storage life of polyester and other thermosetting resins that have a short shelf life.

Initiator A substance that is used to start a chemical reaction and is no longer needed once the reaction is under way.

Injection Molded Card *See Molded Circuit Boards.*

Injection Molding A molding process in which a heat-softened plastic material is forced or injected from a cylinder into a cavity to give the article the desired shape. This process is used with all thermoplastic and some thermosetting molding materials.

Ink A screen-printable thick-film material or paste consisting of glass frit, metals or metal oxides, and solvents.

Ink Blending *See Blending.*

In-Line Package A rectangular package having one row, or two or more parallel rows, of the terminals oriented perpendicular to the package seating plane. *Also called single in-line package.*

Inner Lead Bond (ILB) In tape-automated bonding, the connections made between the chip and the etched conductors on the tape. This is usually a gang bond type of connection. *See also Outer Lead Bond, Tape-Automated Bonding.*

Inorganic Chemicals Chemicals whose structure is based on atoms other than the carbon atom. *See also Organic.*

Inorganic Pigments Carbon-free natural or synthetic salts used for coloring plastics. These pigments are noted for their color stability and weather resistance. Examples are titanium white and chrome yellow. *See also Organic Pigments.*

In Process Any point in manufacturing operations before final testing.

Input A signal that is applied to a circuit.

Input Impedance The impedance presented by the device to the source.

Input/Output (I/O) Pertaining to devices that are used to interface between parts of an electronic assembly. Most commonly used to describe the pins of a semiconductor package. *See also Semiconductor Package.*

Insert The part of a connector that holds the contacts in their proper arrangement and insulates them from one another and from the shell.

Insert Arrangement The number, spacing, and arrangement of contacts in a connector.

Insert Cavity A hole in the connector insert into which the contacts are placed.

Insertion The placing of the pin or lead terminations of components into the holes of a printed wiring board or substrate.

Insertion Force The force required to engage mating components, usually measured in ounces. *See also Engaging and Separating Force.*

Insertion Loss The difference between the power received at the load before and after the insertion of a component, connector, or device at some point in the line.

Insertion-Mount Component (IMC) A leaded component specifically designed for automatic or semiautomatic mounting on printed wiring boards. The leads are inserted in the holes of the board, to a controlled component height above the surface of the printed wiring board, and subsequently soldered to the circuitry on the board surfaces. *See also Surface-Mount Component.*

Insertion-Mount Technology (IMT) As opposed to surface-mount technology, the mounting of leaded components, dual in-line packages, and other leaded electronic packages into through-hole circuit boards. *See also Surface-Mount Technology.*

Insertion Tool A small hand tool used to insert contacts into the connector.

Insert Retention The force applied in either a push or pull direction that an insert is required to withstand without being removed from the connector shell.

Inspection Hole A hole located at one end of a barrel to allow visual inspection to ensure that the conductor has been inserted to the proper depth in the barrel before crimping.

Instruction In a programming language, a meaningful expression that specifies one operation and identifies any operands.

Insulated Gate Field-Effect Transistor (IGFET) A field-effect transistor having one or more gate electrodes that are electrically insulated from the channel.

Insulated Metal Substrate (IMS) A metal substrate that has been coated with an insulating material. Typical examples are porcelainized steel substrates and epoxy-coated metal substrates. These substrates are more dimensionally stable then conventional plastic substrates such as epoxy-glass. They also provide better thermal dissipation and electrical shielding properties. *Also called metal core substrates. See also Porcelainized Steel Substrates.*

Insulated Metal Substrate Technology (IMST) Pertaining to a substrate having a metal core coated with a dielectric such as porcelain or epoxy and having etched circuitry on one or both sides. *Also called metal core substrate technology.*

Insulated Terminal A solderless terminal having an insulated sleeve over the barrel to prevent a short circuit during installation.

Insulating Layer (1) In hybrids, a thick- or thin-film deposited layer of material separating or covering conductive layers. These materials have surface resistivity of 1×10^{13} ohms/sq or volume resistivity of 1×10^{11} ohm - cm. *Also called dielectric layer.* (2) In semiconductors, any thin insulating layer used for separating conductive pads or layers.

Insulation (1) A nonconductive material with a high resistance to the flow of electric current. *Also called dielectric.* (2) A material with low thermal conductivity, that can be used to contain heat within desired boundaries.

Insulation Crimp The area of a terminal, splice, or contact that is formed around the insulation of a wire.

Insulation Displacement Connector A connector capable of making many electrical contacts simultaneously through the insulation on a flat cable.

Insulation Grip A crimp-type contact with an extended cylinder at the rear to accept the base wire and a short length of its insulation. The wire and insulation are held firmly in place after crimping.

Insulation Piercing A crimping method in which lances pierce the wire insulation and make contact with the conductor stands without stripping the insulation off the wire.

Insulation Piercing Terminal A crimping device that pierces wire insulation and makes electrical contact without stripping the insulation from the wires.

Insulation Resistance (1) The electrical resistance of the insulating material between any pair of contacts, conductors, or grounding devices in various combinations. *See also Surface Insulation Resistance.* (2) The ratio of the applied voltage to the total current between two electrodes in contact with a specific insulator.

Insulation System The sum total of the insulation materials used in a system to electrically insulate all electronic components.

Insulator A nonconducting material with a high resistivity used to support or separate conductors. Made from materials with a volume resistivity greater than 10^5 ohms/cm.

Integral Resistor A resistor fabricated within the hybrid substrate pattern. *Also called thick-film resistor. See also Thick Film.*

Integrated Circuit (IC) A small chip of semiconductor material containing an array of active and/or passive components that are interconnected to form a functioning circuit.

Integrated Circuit Package *See Semiconductor Package.*

Integrated Injection Logic Integrated-circuit logic that uses bipolar transistor gates.

Interchip Wiring The conducting paths connecting circuits on one chip with those on other chips in order to complete the electrical circuits.

Interconnection The conductive path required to achieve connection from one circuit element to another or to the rest of the circuit system. Such interconnections may be pins, terminals, formed conductors, soldered joints, wires, or any other mating system.

Interface The junction point or surface between two different or identical materials.

Interface Connection *See Eyelets, Feedthrough, Plated Through Hole.*

Interfacial Bond An electrical connection between the circuitry on the two sides of a substrate.

Interfacial Connection *See Eyelets, Feedthrough, Plated Through Hole.*

Interfacial Gap The distance between the faces of a mated connector.

Interfacial Junction The contact surfaces of the two faces of a connector.

Interfacial Seal A seal that is formed around each contact when the two mating halves of a connector are in compression. The seal is accomplished using an elastomeric material to cover the entire interface area of the mating connective halves.

Interference Protection The measures taken to shield sensitive areas of electrical equipment from electromagnetic and radio frequency interferences.

Interlayer Connection An electrical connection between circuitry in different layers of a multilayer printed wiring board. *See also Buried Via, Hidden Via, Through Connection, Via.*

Interline Dielectric A deposited insulating layer that fills the surface spaces between conductor lines on a substrate and matches them in height.

Intermateable Commonly used to describe certain interchangeable connector halves which are made by various manufacturers.

Intermetallic Bond The ohmic contact established when two metal conductors are welded or fused together.

Internal Inspection *See Preseal Visual Inspection.*

Internal Layer A conductive layer that is contained entirely within a multilayer printed circuit board.

Internal Resistance The thermal resistance encountered between the junction of two dissimilar materials inside a package to a point on the outside surface of the package. *See also Thermal Resistance.*

Internal Visual *See Preseal Visual Inspection.*

International Annealed Copper Standard (IACS) The conductivity of annealed, unalloyed copper, arbitrarily rated as 100 percent IACS. The conductivities of all other metals and alloys are expressed as a percentage of this standard.

Interpenetrating Polymer Network (IPN) Two or more polymers that have been reacted together so that the polymers penetrate one another in the final polymer form. A combination of good properties is achieved in this process.

Interstice (1) A very small space between two objects that are purposely placed in close proximity. (2) A void or valley between individual strands in a conductor or between insulated conductors in a multiconductor cable.

Interstitial Via A plated through hole connecting two or more conductive layers of a multilayer printed circuit board. *See also Blind Via, Buried Via.*

Intraconnection The joining of conductors in a circuit on the same substrate.

Intrinsic Conduction A process of electrical conduction in a pure or intrinsic semiconductor material that contains no impurities. *See also Intrinsic Semiconductor.*

Intrinsic Dielectric Strength The characteristic dielectric strength of a material under ideal conditions. *See also Dielectric Strength.*

Intrinsic Semiconductor A semiconductor whose charge-carrier concentration is essentially the same as that of an ideal semiconductor crystal. *See also Extrinsic Semiconductor, Semiconductor.*

Invar A nickel-iron alloy having a very low coefficient of thermal expansion. *See also Coefficient of Thermal Expansion, Kovar.*

I/O *See Input/Output.*

Ion An electrically charged atom or group of atoms. Positively charged ions have a deficiency of electrons, while negatively charged ions have a surplus of electrons. *See also Anion, Cation, Electron.*

Ion Exchange A reversible chemical reaction between a solid and a fluid by means of which ions are interchanged from one substrate to another.

Ion Exchange Resin Small particles containing acidic or basic groups that will trade ions with salts in solutions.

Ionic Activity A measure of the pH of a solution. *See also pH.*

Ionic Cleanliness The degree of surface cleanliness with respect to the number of ions or weight of ionic matter per unit of square surface. *See also Ionic Contaminant.*

Ionic Contaminant Any polar (ionic) contaminant that, when dissolved in water as free ions, increases electrical conductivity. Typical contaminants include flux activators, fingerprints, and plating salts. *See also Ionograph.*

Ionic Solvent *See Polar Solvent.*

Ion Implantation A method of doping semiconductors in which the desired dopant is ionized and accelerated by an electric field, penetrates the surface, and is deposited within the semiconductor material.

Ionizable Material A material whose electrons are readily lost from atoms or molecules, thereby reducing the electrical resistance of the material.

Ionization A process by which electrons are lost from or transferred to neutral molecules or atoms to form positively or negatively charged particles.

Ion Migration The movement of free ions within a material or across the boundary of two materials under the influence of an applied electric field. *See also Electrostatic Shield.*

Ion Milling A dry etching operation performed by ions, usually in a gaseous form, on a very large scale integration basis. The results are similar to those achieved with wet chemical and plasma etching.

Ionograph An instrument for measuring the electrical conductivity of the solution that has been used to clean circuit board assemblies, in order to determine the ionic contamination level. *See also Ionic Contaminant.*

Irradiance The radiated power per unit area incident upon a surface, measured in watts/cm^2.

Irradiation In insulations, the exposure of material to high-energy emissions for the purpose of favorably altering the molecular structure by crosslinking.

Isolation The positioning or shielding of an electrical component or circuit so that it can withstand interferences. *See also Electromagnetic Shield.*

Isostatic Press A hydraulic press that applies equal pressure in all directions.

Isotactic The structure of molecules that are polymerized in parallel arrangements of radicals on one side of the carbon chain.

Isotropic Pertaining to a material whose electrical and optical properties are the same in every direction in the material. *See also Anisotropic.*

Izod Impact Strength A measure of the toughness of a material under impact. *See also Izod Impact Test.*

Izod Impact Test A test for measuring the impact resistance of a plastic material. A notched sample bar of plastic is struck by a pendulum, and the force required to break the sample is a measure of the impact strength.

Jacket A plastic, rubber, or synthetic covering over the insulation, core, or sheath of a cable.

Jackscrew A screw attached to one half of a two piece contact that is used to pull both halves together or to separate them.

Jet Impingement Cooling Liquid cooling of microelectronic devices achieved by passing a coolant through a capillary tube or orifice whose jet stream is pointed so as to impinge on the surface to be cooled. *See also Microchannel Cooling.* See Fig. 8.

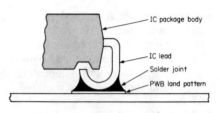

Figure 9: J-Lead

Joint The location at which two adherends are held together with a layer of adhesive.

Joint Army-Navy (JAN) A government trademark identification used as the part number prefix for most devices procured to the requirements of the military specifications and standards.

Josephson Junction The close proximity of two superconducting materials that are separated by a thin insulating barrier to act as an electric switch.

Josephson Superconducting Device Two superconductive ceramic materials that act as a Josephson device at low temperatures.

Joule The work done by a force of one newton acting through a distance of one meter.

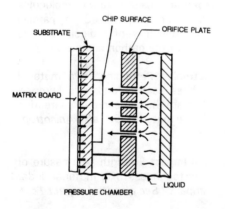

Figure 8: Jet Impingement Cooling

J Lead A package lead that has been shaped to resemble the letter J. *See also Gull Wing Lead.* See Fig. 9.

Jumper Cable A short cable used to interconnect two printed wiring boards or other assemblies.

Jumper Wire (1) A direct electrical connection between two points on a printed wiring board or substrate that is used to complete a circuit temporarily or to bypass a circuit. (2) An electrical connection that is not a part of the original design. *Also called lead wire.*

Junction (1) A point in a circuit where two or more wires are connected. (2) A contact between two dissimilar metals or materials. (3) A region of transition between P-type and N-type semiconductive materials in transistors and diodes.

Junction Gate Field-Effect Transistor (JFET) A field-effect transistor having one or more gates that form P-N junctions with the channel.

Junction Temperature The temperature of a semiconductor junction.

Just in Time (JIT) Production control techniques that minimize inventory by delivering parts and materials to a manufacturing facility just before they are incorporated into a product.

K (1) The symbol for dielectric constant. (2) Abbreviation for kilo (1000). (3) A 1K-memory chip, which actually contains 1024 bits because it is a binary device based on powers of 2. (4) The symbol for thermal conductivity.

Kapton A high temperature polyimide film or tape made by DuPont.

Kerf A slit or channel cut in a thick-film resistor by laser trimming or by abrasive trimming. *Also called notch.*

Kevlar The trademark name for a group of DuPont aromatic poly-imides that are frequently used as fibers in reinforced plastics and composites. Major characteristics are low thermal expansion, light weight, and good electrical properties, coupled with stiffness in laminated form. One special application is in high-performance circuit boards requiring low *X-Y* axis expansion.

Key A short, hardened pin that slides into a corresponding mating hole to guide the two mating halves in the engagement and assembly of a connector.

Keying A mechanical arrangement of guide pins, sockets, plugs, bosses, slots, and grooves in a connector housing, shell, or insert. The arrangement allows connectors of the same size and type to be properly aligned without the danger of making a wrong connection and damaging the mating pins and sockets.

Keying Slot A slot in a printed board that allows only one possible way of mating with a connector.

Keyway A groove or slot that ensures the correct orientation and subsequent mating of two connector halves.

K Factor (1) The symbol for thermal conductivity. (2) The amount of heat that passes through a unit cube of material in a given time when the difference in temperature of the two faces is 1°C.

Kiln A high-temperature furnace used for firing ceramic materials.

Kilo- A prefix indicating a multiple of one thousand.

Kilohertz (kHz) Thousands of cycles per second. *See also Gigahertz, Megahertz.*

Kirkendahl Voids Voids that are induced at the interface between two different metals with different interdiffusion coefficients.

Known Good Die (KGD) A bare chip semiconductor die that has been reliably tested to predetermined specifications so as to ensure high quality.

Kovar A nickel-iron-cobalt alloy. Initially a Westinghouse trade name. This class of alloys is used in microelectronic and hybrid packages because of its low coefficient of thermal expansion. The thermal expansion of Kovar closely matches that of glass compositions used in glass-to-metal seals. *See also Invar, Alloy 42.*

Kovar Tab *See Moly Tab.*

Lacing Cord A flexible, flat cord, available in several different materials, that is either coated or impregnated. It is used for tying cable forms, hookup wires, bundles of wires, and wire harness assemblies.

Laminae A set of single plies or layers of a laminate (plural of *lamina*).

Laminar Flow A directed flow of filtered air that is moved constantly across a clean work area in a direction parallel to the surface of the station. Laminar flow is also a factor in flow of heat and thermal management. *See also Turbulent Flow.*

Laminar Wave Soldering A solder wave having little or no surface turbulence.

Laminate (1) To bond sheets of material together, usually with heat and pressure. (2) Sheets of material that are bonded together. A common laminate form is composed of fabric sheets that are wetted by resin and bonded together, such as epoxy-glass circuit boards. These laminate sheets may have a metal foil on one or both surfaces. Metal sheets bonded together, such as copper-invar-copper, are called *metal-clad laminates.*

Laminated Ceramic Package A type of electronic package consisting of a multilayer cofired ceramic body (usually with a cavity to accept the chip) brazed leads or leadless solder pads, and a hermetically sealed metal lid. *See also Cofired Ceramic.*

Laminated Plastics Layers of resin-impregnated reinforcements bonded together by heat and pressure to form a single sheet. Resins include phenolics, melamines, epoxies, silicones, polyesters, and polyimides.

Laminate Void Absence of resin in any cross-sectional area that should normally contain resin.

Land A widened area of a conductor on a substrate that is used for the attachment of wire bonds or mounting and bonding of chip devices. *Also called bonding area, bond pad, or footprint.*

Landless Hole A plated through hole without any land or conductive material surrounding the hole.

Land Pattern A group of lands or footprints that are used for chip mounting, wire bonds, or testing. *See also Footprint.*

Lanyard A device used to uncouple connector halves by pulling a cord or wire. Only certain connectors are equipped with this device.

Lap Joint A joint made by placing one adherend partly over another and bonding the overlapped portions. *See also Adherend.*

Lapping A grinding and polishing process used to obtain a precise thickness or an extremely flat, highly polished surface.

Lap Shear Strength The shearing pressure at which an adhesive-bonded and cured lap joint fails. *See also Shear Strength.*

Large-Scale Integration (LSI) An array of integrated circuits on a single substrate that usually contains more than 100 gates or circuits and at least 500 elements.

111

See also Medium-Scale Integration, Small-Scale Integration.

Laser (1) *L*ight *A*mplification by *S*timulated *E*mission of *R*adiation. (2) A device that, when stimulated by a signal, produces a light beam by the emission of energy stored in a molecular or atomic system.

Laser Bonding A process in which a metal-to-metal bond is formed by the use of a laser beam as the heat source.

Laser Pattern Generation A method in which a precision laser beam is used as the light source to produce photo images for original artwork or direct pattern generation for integrated and other electrical circuits.

Laser Soldering A selective soldering process that uses a programmable laser system for soldering applications.

Laser Trimming The increasing of base thick-film resistor values by using a focused laser beam to cut and vaporize the resistor material.

Laser Welding A welding process in which the thermal energy is supplied by a laser impinged on the surface of the metal to be welded.

Latch-Up A condition in which the output of a circuit has become fixed near one of the two voltage extremes and the input is no longer able to exert effective control.

Latent Curing Agent A curing agent that produces long-time stability at room temperature but rapid cure at an elevated temperature.

Lattice Structure The stable arrangement of atoms and their electron-pair bonds in a crystal.

Lay The axial distance for a single strand of a stranded conductor or an insulated conductor of a cable to complete one turn around the axis of the conductor or cable.

Lay Direction The twist in the top strands of a cable as viewed along the axis of the cable away from the observer. The twist can be a right-hand or a left-hand lay.

Layer A single film of material in a multilayer structure of a substrate or laminate.

Layout The positioning of conductors and resistors on artwork prior to photoreduction to obtain a negative or positive used in subsequent fabrication.

Lay-Up The process of registering and stacking individual layers of a multilayer board in preparation for subsequent fabrication steps in laminating or vacuum bag molding. *See also Laminate, Vacuum Bag Molding.*

L Cut A trim notch cut in a thick-film resistor. The cut starts perpendicular to the resistor length and then turns 90° to finish the trim parallel to the resistor axis.

Leaching The dissolving or alloying of a metal to be soldered into the molten solder. Leaching of gold plating or films into solders is a common problem and leads to poor solder joints.

Lead A wire that connects two points in a circuit and that is usually self-supporting.

Leaded Pertaining to electronic devices that have electrical leads extending from their enclosures.

Leaded Ceramic Chip Carrier (LDCC) A square or rectangular ceramic form with a cavity to house a semiconductor device and with high-density input/output (I/O) pads and leads on two or four sides of the package. *See also Chip Carrier.*

Leaded Surface-Mount Component *See Surface-Mount Component.*

Leadframe A rectangular metal frame with leads. The frame contains the leads, which are connected to semiconductor dies. After encapsulation or lidding of the package, the frame is cut off, leaving the leads extended from the package.

Leadless Pertaining to electronic devices that do not have electrical leads extending from their enclosures but rather solder lands or bumps located on the top, bottom, or sides of the package.

Leadless Chip Carrier (LCC) A square or rectangular ceramic package with a cavity to house a semiconductor device or other electronic components. It has metallized

interconnecting lands instead of leads. These packages are reflow-soldered to a printed board or substrate. *See also Chip Carrier.* See Fig. 10.

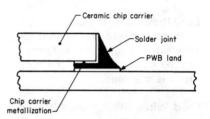

Figure 10: Leadless Chip Carrier

Leadless Device A device having metallized interconnecting lands instead of leads. Chip capacitors and other components are most typical.

Leadless Inverted Device (LID) A square or rectangular ceramic carrier with a cavity used to house a semiconductor chip. It has metallized conductor lands for interconnections to the semiconductor device and metallized pedestals for reflow soldering to a printed board or substrate. The carrier is normally coated with a plastic to protect the bonds and chip device and is mounted in an inverted position on the printed board or substrate.

Leadless Multiple Chip Hybrid (LMCH) A hybrid microcircuit device in a leadless chip carrier package. *Also called leadless hybrid device (LHD).*

Leadless Pad Grid Array (LPGA) A leadless surface-mount package containing input/output (I/O) connections through metallized pads arrayed in a matrix on its base. Used primarily for integrated circuits with high pin counts.

Lead Projection The distance that a component lead protrudes through the side of a printed board opposite to the side upon which the component is mounted.

Lead Wires Wires used for intra-connections or input/output (I/O) leads. *Also called jumper wires.*

Leakage Current A small, stray, undesirable amount of current that flows through or across an insulator between two electrodes.

Leak Check A method for testing the hermeticity of an electronic package. Two types of tests are used: fine leak and gross leak. *See also Fine Leak, Gross Leak.*

Leak Detectors Devices used to detect sealing defects in hermetically sealed packages. These are very fine leaks that cannot be detected by gross-leak methods. The packages are placed in a bomb and pressurized with helium; a helium spectrometer is then used to detect and locate the leak. *See also Fine Leak, Gross Leak.*

Legend A format of letters, numbers, symbols, and patterns used primarily to identify component locations and orientations for con-venience of assembly and maintenance operations.

Leno Weave A locking-type weave in which two or more warp threads cross over one another and interlace with one or more filling threads. It is used primarily to prevent shifting of fibers in open fabrics.

Leveling In thick-film technology, the smoothing out or self-leveling of screen mesh marks of pastes after a pattern or circuit has been printed.

Levels of Electronic Packaging and Interconnecting There are four levels of interconnection in electronic packaging: (1) between the device and the package; (2) from package to package; (3) from printed wiring board to printed wiring board, such as mother board to line-replaceable unit (LRU); and (4) between the LRU and the system.

Lid A flat piece of metal or ceramic used to seal electronic packages. Stepped or depressed lids have a flange to conform to the inside of the package.

Lid Hole A purposely made pin-size opening in the middle or the corner of a lid or cover of and electronic package. It may be either a burr-free prestamped hole serving as the vapor inlet during the Parylene coating of sealed hybrid units or a drilled lid hole on sealed hybrid units. The hole is normally covered with sticky tape, which captures loose particles during particle impact noise detection (PIND) vibration testing. The hole is subsequently closed off by solder seal. *See also Particle Impact Noise Detection, Parylene.*

Life Aging A burn-in test conducted at an elevated temperature for extended periods of time to assess the quality of the product. *See also Burn-In.*

Life Cycle Test A test designed to indicate the time span of a device before failure. It is conducted in a controlled, accelerated environment. *See also Accelerated Aging.*

Life Drift A permanent change in the value of a device or level of a circuit element under load. It is rated as a percentage change from the original value for 1000 hours of life.

Life Test The test of a component under conditions similar to those under which the component operates, or conditions simulated by acceleration, in order to determine the life expectancy of the component. *See also Accelerated Aging.*

Lift-Off Mark The image of the bond area that remains after the bond has been removed.

Light-Emitting Diode (LED) A P-type or N-type semiconductor device specifically designed to emit light when forward-biased.

Lightning Electromagnetic Pulse (LEMP) A disturbance created by the electromagnetic field from lightning.

Limited-Coordination Specification or Standard A specification or standard that has not been fully coordinated with or accepted by all interested purchasers or programs. It applies primarily to military agency documents and is issued to cover requirements unique to one particular program.

Linear A straight-chain organic molecule with linear carbon-to-carbon linkages. Branching often exists off the linear chain. *Also called aliphatic. See also Hydrocarbon.*

Linear Coefficient of Expansion *See Coefficient of Thermal Expansion.*

Linear Integrated Circuit (LIC) A circuit in which the output voltage, usually within a limited range of signal voltages and frequencies.

Linear Microcircuit *See Linear Integrated Circuit.*

Line Definition In thick-film technology, the degree of sharpness or clarity of screen-printed lines.

Line Discontinuity A point on a transmission line equivalent to a separate circuit having resistance, capacitance, and inductance. The discontinuity produces false reflections.

Line Impedance The impedance as measured across the terminals of a transmission line.

Line Loading The connecting of external resistance, inductance, and capacitance in a transmission line.

Line-Replaceable Unit (LRU) An electronic subassembly, equipped with a multipin connector, that can readily be removed and replaced with another, identical subassembly.

Line Resistance The resistance offered by conductor lines in a package. It is measured in ohms per unit length or for a given cross section in ohms per square. *See also Sheet Resistivity.*

Lines Conductor runs of an interconnection network.

Lines per Channel The number of conductive lines between metallized holes or lands in a printed wiring board or a ceramic substrate.

Liquid Crystal Display (LCD) A visual display in which a liquid crystal material is hermetically sealed between glass plates. The application of a magnetic field or of voltage or heat alters the crystalline properties, rendering segments visible by color contrast.

Liquid Crystal Polymers (LCPs) Polymers that spontaneously order themselves in the melt, allowing relatively easy processing at relatively high temperatures and resulting in excellent preparation for electronic systems. They are characterized as rigid rods. Kevlar and Nomex are examples, as are numer-

ous high-performance thermoplastic resins used in electronics.

Liquid Injection Molding A fabrication process in which catalyzed resin is metered into closed molds.

Liquidus The temperature of a metal or alloy at which it is completely liquid.

Lithography *See Photolithography.*

Live-Line Connector *See Hot-Line Connector.*

Loadbreak Connector A connector that is designed to open and close while current is flowing through the circuits; similar to a relay.

Load Life That period of time over which a component can sustain its full power rating.

Locating Notch A tooling feature in the form of a notch in a printed board.

Locator *See Stop Plate.*

Locking Spring *See Contact Retainer.*

Logic Array *See Gate Array.*

Logic Design The selection and interconnection of logic units to achieve logical function in the digital system.

Logic Gate A combinational logic function consisting of an unspecified number of inputs and outputs and performing one of the Boolean functions (and, or, nand, nor, or inversion) or transmission.

Logic Microcircuit *See Digital Integrated Circuit.*

Logic Part A group of logic circuits that are interconnected and packaged and used throughout a digital system.

Logic Primitive A basic logic function as it exists as a single unit.

Logic Service Terminal (LST) A terminal that carries logic signals.

Longitudinal Indent An indent whose longest dimension is in line with the barrel of a connector.

Loop The curvature of the wire between each end of the two attached wire bonds.

Loop Height The maximum perpendicular distance from the top of the wire loop to a point between the two attached wire bonds.

Loss A decrease in power in the transmission of a signal from one point to another.

Loss Angle *See Dissipation Factor.*

Loss Factor The rate at which heat is generated in a insulating material. It is equal to its dissipation times the dielectric constant. *See also Dissipation Factor.*

Loss Tangent *See Dissipation Factor.*

Lossy Signal Line A transmission line containing series resistance, skin effect, and dielectric conduction.

Low Frequency In the radio frequency spectrum, the band from 30 to 300 kHz.

Low-Level Noise Tolerance The maximum noise level in the receiver when the input signal is in its down state.

Low-Loss Dielectric An insulating material that has a low power loss over long lengths, making it suitable for transmission lines. It usually also has a low dielectric constant and low dissipation factor.

Low-Loss Substrate A substrate with high radio frequency resistance and low energy absorption. It usually also has a low dielectric constant and low dissipation factor.

Low-Pressure Laminates Laminates that can be molded and cured at 400 psi or below.

Low-Solids Solder Flux *See Flux, No-Clean.*

Low-Stress Polyimides Polyimides whose polymerization stress is lowered by virtue of thermal expansion and/or crystalline molecular structure. This low-stress property results in reduced damage to coated semiconductors. *See also Liquid Crystal Polymers, Planar, and Polyimide.*

Low-Surface-Energy Materials Materials that have surface-free energies below 500 ergs/cm^2 and low melting points. They are not easily wetted or bonded. *See also High-Surface-Energy Materials, Nonwetting Surface.*

Low-Temperature-Firing Ceramic A ceramic composition that can be fired at temperatures considerably below those used in normal kiln firing such as for pressed ceramics. These compositions are often glass-ceramic. *See also Kiln, Pressed Ceramic.*

Lug A device that can be soldered or crimped to the end of a wire and has a eye or fork for mechanical attachment.

Lumped Element In microwave electronic systems, a combination of elements or electrical functions in a single discrete part. *See also Distributed Element.*

Macerate To chop or shred fabric for use as a filler for a molding resin.

Macrostructure The structure of a properly prepared specimen as seen at low magnification, such as with the naked eye, and up to 50X.

Magnet Wire Enamel-coated copper wire chiefly used for winding electrical coils. In hybrid technology, it is used as haywire or jumper wire in electronic modules. *See also Haywire, Jumper Wire.*

Mainframe *See Central Processor.*

Maintainability The ability of a part, subsystem, or system to continue to operate, or to be restored to a specified level of operation, when the maintenance is performed in accordance with recommended procedures and resources.

Major Defect A defect that is likely to result in a failure of a unit or product, thereby materially reducing its usability for its intended purpose.

Major Weave Direction The continuous-length direction of a roll of woven-glass fabric.

Mandrel A form around which resin-impregnated fiber is wound to form pipes, tubes, or vessels by the filament-winding process.

Manhattan Effect *See Tombstoning.*

Manhattan Length The wire length between terminals of a net or connection measured in the x and y directions on the wiring plane of a package.

Manufacture To make or produce by hand or by machinery, on a small or large scale, a usable product or part.

Margin The distance between the reference edge of a flat cable and the nearest edge of the first conductor.

Mask A patterned screen of any of several materials and types used for (1) shielding and exposing selected areas of a semiconductor or substrate during its processing or (2) shielding and exposing selected areas of a photosensitive surface on the substrate to a light pattern so that the photosensitive surface itself becomes a mask. The mask can be designated either by type (e.g., oxide mask or metal mask) or by function (e.g., diffusion mask or vapor deposition mask).

Mass Spectrometer An instrument capable of making rapid analysis of chemical compounds by means of gas ionization.

Mass Splice In fiber optics, a technique in which all the fibers are joined simultaneously without handling each fiber individually.

Mass Termination A termination process in which terminals pierce flat cable insulation without stripping the insulation, and mate by cold flow with the conductors to form a metal-to-metal joint.

Master Artwork *See Master Layout.*

Master Batch Principle A blending of resistor pastes used in thick-film technology to a nominal value of ohms per square. The nominal value is then the master control number.

Master Drawing A drawing that shows the dimensional limits to or grid locations of any and all parts of a printed circuit. These limits or locations include size, type, position of holes, and other pertinent information necessary to fabricate the product.

Master Layout The original layout of a circuit.

Master Pattern A one-to-one scale pattern used to produce printed circuit boards within the accuracy of the master drawing. *Also called production master.*

Master Slice A silicon wafer containing many clusters of components. These elements can be interconnected with metallization paths to form the desired circuits. The wafer is then diced to form single circuits. *See also Wafer.*

Mat A randomly distributed felt of glass fibers used in reinforced plastics lay-up molding or in laminates.

Matched Metal Molding A method of molding reinforced plastics between two close-fitting metal molds mounted in a hydraulic press.

Matched Seal A glass-to-metal seal in which Kovar and glass, both having the same coefficient of thermal expansion, are used to form hermetic seals. *Also called glass-to-metal seal. See also Invar, Kovar.*

Matching (1) An indication of how closely two resistors, or capacitors, in a network approximate each other's value. Matching is significant in resistor-ladder networks to ensure that the resistors are within a given ratio, expressed as a percentage of nominal value. (2) Selection of two or more parts to provide characteristics when used together that similar parts selected at random would not provide.

Mate The joining of two connector halves.

Material Density The weight per unit volume of any given material.

Matrix (1) An orderly two-dimensional array with circuit elements such as wires, diodes, and relays arranged in rows and columns. (2) The medium to which something is added to achieve special improved properties. For example, epoxy resin is the matrix in glass-epoxy laminates and aluminum is the matrix in most metal matrix composites. *See also Ceramic Matrix Composites, Composite, Metal Matrix Composites, Organic Composites.*

Matrix Tray *See Waffle Pack.*

Matte Finish A surface finish having a grainy appearance.

Maximum Rating A rating that establishes either a limiting capability or a limiting condition beyond which damage to the device may occur. A limiting condition may be either a maximum or a minimum. *See also Rating.*

Mealing A separation of a conformal coating from the base material on a printed board. It usually appears in the form of spots or patches and is observable as blisters under the coating. *Also called vesication.*

Mean Time Between Failures (MTBF) The average time, expressed in hours, between failures of a device, circuit, or system on a continuous operating basis. Used to express reliability level.

Mean Time to Failure (MTTF) The average of the lengths of time to failure for parts of the same type that are operated as a group under the same conditions.

Measling A condition that occurs in laminated base material in which internal glass fibers are separated from the resin at the weave intersection. This condition manifests itself in the form of discrete white spots or crosses that are below the surface of the base material. It usually is caused by thermally induced stress. *See also Crazing.*

Mechanical Pertaining to machinery or tools.

Mechanical Properties Material properties associated with elastic and inelastic reactions to an applied force.

Mechanical Wrapping The physical securing of a wire lead or component lead around a solder terminal before soldering.

Medium-Scale Integration (MSI) Integrated circuits that have fewer than 100 gates or basic circuits but at least 12 equivalent gates or 100 circuit elements. *See also Large-Scale Integration, Small-Scale Integration.*

Mega- A prefix indicating a multiple of one million.

Megahertz (mHz) Millions of cycles per seconds. *See also Gigahertz, Kilohertz.*

Melamines Thermosetting resins made from melamine and formaldehyde that possess excellent hardness, clarity, and electrical properties.

Melt Molten plastic, in the melt phase of the plastic material during the molding cycle.

Melting Time The period of current flow, after circuit protector activation, required to melt a fuse element. *See also Clearing Time.*

Memory A component that provides ready access to data or instructions previously recorded, such as in a computer or control system, so as to make them respond to a problem.

Memory Adder A device added to the basic cycle-per-instruction rate of the central processor.

Memory Chip A semiconductor device that stores information for later use.

Meniscugraph Test A solderability test that records surface tension of the solder bath. The test specimen is connected to equipment through a strain gauge, which records the surface tension of the solder bath.

Mesh Size The number of openings or squares per linear inch in a screen. A 100-mesh screen has 100 openings per linear inch.

Metal A material that has high electrical and thermal conductivity at room temperature and exhibits malleability and ductility characteristics.

Metal-Clad Base Material Same as metal-clad laminate, except that the clad material may be a homogeneous material rather than a laminated material. *See also Laminate, Soft Substrate.*

Metal-Clad Laminate *See Laminate.*

Metal Core Substrate Technology *See Insulated Metal Substrate Technology.*

Metal Electrode Face (MELF) A tubular-shaped component with metallized terminations for surface mounting.

Metal Inclusion A metal particle embedded in a nonmetal material.

Metal-Insulator-Semiconductor (MIS) Technology The technology whereby circuits and circuit elements are formed with the use of a three-layered semiconductor-insulator-metal structure. Field-effect transistors, capacitors, nonlinear resistors, variable-threshold diodes, and similar circuit elements can be formed with this technology.

Metal Laminate *See Laminate.*

Metallization (1) The network of conductive material formed on the surface of a chip or substrate to interconnect microcircuit elements. *See also Bottom Metallization, Multilayer Metallization.* (2) A conductive film, in single or multiple layers, deposited on the surface of another material, usually by either vapor or chemical action, to perform electrical and mechanical functions.

Metallized Ceramic (MC) A ceramic substrate that has been metallized with screened thick films, or evaporated thin films, or by some other process.

Metal Mask A thin sheet of metal in which holes are etched to a desired pattern. Used for precision and/or fine printing and for solder cream printing.

Metal Matrix Composites A formed metal part, which is the matrix, reinforced with certain fibers or flakes to achieve special properties. Aluminum is most commonly the matrix, which, when reinforced, is stronger than nonreinforced aluminum. Other metal matrix materials include copper and titanium. Reinforcements include silicon carbide and graphite. *See also Ceramic Matrix Composites, Matrix.*

Metal-Nitride-Semiconductor (MNS) Technology A subcategory of metal-insulator-semiconductor technology, in which the insulator employed is a nitride. *See also Metal-Insulator-Semiconductor Technology.*

Metal-Oxide-Semiconductor Field-Effect Transistor (MOSFET) An insulated gate field-effect transistor in which the insulating layer between each gate electrode and the channel is oxide material.

Metal-Oxide-Semiconductor (MOS) Technology A subcategory of metal-insulator-semiconductor technology, in which the insulator employed is an oxide of the semiconductor substrate material. The term MOS is often misused to include other categories of insulated gate technology such as MIS, MNS, and silicon gate (SIS). *See also Metal-Insulator-Semiconductor Technology.*

Metal Thermal Via *See Through Via.*

Metal Thick-Nitride Semiconductor (MTNS) A semiconductor device with a thick silicon nitride or silicon nitride-oxide layer used in place of an oxide as a passivation coating. *See also Passivation.*

Metal Thick-Oxide Semiconductor (MTOS) A semiconductor device with has a thicker oxide layer outside the active gate area to reduce parasitic problems. *See also Passivation.*

Metal-to-Glass Seal *See Glass-to-Metal Seal.*

Microbond An interconnecting bond using a small-diameter (0.001 inch), usually gold wire, to a conductor or to a die.

Microchannel Cooling Liquid cooling of microelectronic substrate or devices achieved by passing a coolant through small round or rectangular channels, which are formed in the substrate or device. *See also Jet Impingement Cooling.*

Microcircuit A small circuit composed of interconnected elements within a single package having a high equivalent circuit element density and performing an electronic circuit function. Printed wiring boards, circuit card assemblies, and modules composed exclusively of discrete electronic parts are excluded.

Microcircuit Module An assembly of microcircuits, or an assembly of microcircuits and discrete parts, designed to perform one or more electronic circuit functions and so constructed that for the purposes

of specification, testing, commerce, and maintenance, it is considered to be indivisible.

Microcomponents Miniature components such as chip transistors, chip capacitors, and chip resistors.

Microcracks Small, thin cracks that can be seen only with a microscope at magnifications approaching 100X.

Microelectronic Device *See Microcircuit.*

Microelectronic Hybrid Package (MHP) A geometrically shaped enclosure containing a ceramic substrate with thick- or thin-film circuitry, discrete components, and wire bond interconnections. The substrate is bonded to the base of the enclosure, which is equipped with terminals to provide electrical access to the devices inside. A lid or cover is soldered or welded to the top of the enclosure to form a hermetic seal. Through leads interconnect the devices and circuitry inside the package to circuitry outside the package, usually on a printed board.

Microelectronics The area of electronic technology associated with or applied to the realization of electronic systems from extremely small electronic parts or elements.

Microfinish A microinch or micrometer surface finish on a material. The finish is uniform in thickness and free of porosity, peaks, and valleys.

Microlithography The photo-lithography of the finest feature electronic devices, including optical, X-ray, and electron beam techniques for high-functioning semiconductor devices. *See also Photolithography, Semiconductor Device, Wafer.*

Microminiaturization The technique of packaging microminiature circuits and miniature components in a high-density assembly through the use of thick-film or thin-film technology or other fine-circuitry fabrication technique. *See also Microelectronic Hybrid Package, Miniaturization, Multichip Module.*

Micron A unit of length equal to 10,000 A, 0.0001 cm, or 0.000039 inch. A practical equivalent is that 1 mil equals approximately 25 microns.

Micropositioner A tool used to accurately position a substrate or a device for subsequent bonding or trimming.

Microprobe A sharp-pointed miniature probe with a positioning handle, used for making temporary electrical contact to a device or circuit for testing. *See also Bed of Nails, Probe.*

Microprocessor An integrated circuit that provides, in a single chip, functions equivalent to those in a central processing unit of a computer.

Microsectioning A series of steps— which include cross-sectioning,

encapsulation, polishing, and etching— used in the preparation of a specimen for microscopic examination.

Microstrip A type of microwave transmission line configuration that consists of a conductor over a parallel ground plane separated by a dielectric. *See also Stripline.* See Fig. 11.

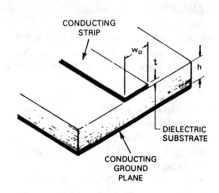

Figure 11: Microstrip

Microstructure (1) The structure of a properly prepared specimen as seen at high magnification. (2) A structure comprised of microsize particles that are bound together, such as the internal crystalline structure of a material. *See also Macrostructure.*

Microwave Wavelengths of 0.3 to 1.0 mm and a frequency range of 1000 to 300000 mHz.

Microwave Integrated Circuit (MIC) A miniature microwave circuit using hybrid circuit technology to form the conductors and attach the chip devices and components. *See also Multichip Hybrid Package.*

Migration An undesirable movement of metal ions, especially silver, from one location to another in the presence of moisture and an electrical potential. The result can be low-resistance paths or shorts between conductors.

Mil (1) A unit of length equal to 0.001 inch or 0.0254 mm. (2) A unit used in measuring the diameter of a wire or thickness of insulation over a conductor: 1 mil = 0.001 inch. *Also called a thou.*

Military With respect to standards and specifications, pertaining to the armed forces of the United States—Army, Navy, Air Force, Marines—or other Department of Defense units.

Miniaturization A technique of packaging by reducing the size or weight of parts and arranging them for maximum utilization of space. *See also Microminiaturization.*

Minimum Annular Ring The minimum width of metal, at the narrowest point between the edge of the hole and the outer edge of the terminal ring.

Minor Defect An irregularity or imperfection that does not impair the usability of a unit for its intended purpose.

Minor Weave Direction The width direction of a roll of woven-glass fabric.

Misalignment Loss In fiber optics, the loss attributed to the lateral or angular misalignment of the optical junction centerline.

Mislocated Bond A wire bond in which part of the bond area is off the bonding pad.

Mixed-Component Mounting Technology An electronic packaging technology that utilizes both surface-mount and through-hole technologies within the same electronic package and interconnection levels. *See also Surface-Mount Technology, Through-Hole Component.*

Mock Leno Weave An open-type weave accomplished by a system of interlacings that draws a group of threads together and leaves a space between the next group. The warp threads do not actually cross one another, as in a real leno weave, so no special attachments are required for the loom. This type of weave is generally used when a high thread count is required for strength but the fabric must still remain porous.

Modem A device that modulates and demodulates signals transmitted over data communication facilities. The word *modem* was derived from modulator-demodulator.

Modifier A chemically inert ingredient, such as a filler, added to a plastic resin formulation to change

125

its properties in some desired characteristic or characteristics. *See also Filler.*

Modular Connector A connector that is capable of having similar or identical sections added onto the original connector to provide additional capabilities.

Modulation The effect of signal fluctuations on the radio frequency carrier.

Module A subassembly in an electronic packaging plan containing components, built to a standard size and equipped with standard plug-in or solder terminations. *See also Standard Module.*

Modulus of Elasticity The ratio of unidirectional stress to the corresponding strain, or the slope of the line, in the linear stress-strain region below the proportional limit. For materials with no linear range, a secant line from the origin to a specified point on the stress-strain curve or a line tangent to the curve at a specified point may be used.

Moisture Absorption The amount of water pickup by a material when that material is exposed to water vapor. Expressed as a percentage of the original weight of the dry material.

Moisture Resistance The ability of a material to resist absorbing moisture, either from the air or when immersed in water.

Moisture Stability The stability of a circuit to function electrically in a high-humidity environment.

Moisture Vapor Transmission The rate at which moisture vapor passes through a material, at specified temperature and humidity levels. Expressed as g/mil of material thickness/24 hr/100 in^2.

Mold (1) A medium or tool designed to form desired shapes and sizes. (2) To process or fabricate a plastic material using a mold.

Molded Formed by pouring or forcing liquid or soft plastic material into a mold cavity. *See also Cast, Compression Molding, Injection Molding, Pot.*

Molded Circuit Boards Printed circuit boards that have been injection-molded for application in three-dimensional (3-D) molded circuits. *See also Molded Circuits.*

Molded Circuits Electrical circuits that have been printed and etched on a flexible release material and subsequently transferred and bonded to a contoured surface such as the inside of a plastic cover or chassis or a molded circuit board. *Also called 3-D circuits,* because they are in three dimensions.

Molded Plastic Semiconductor Packages Electronic packages that are fabricated by transfer molding a leadframe subassembly, with chips and fragile wires, to achieve the lowest possible assembly cost. Transfer molding is most commonly used because of the low viscosity of the resins and the low pressure of the process, which does not damage the fragile assembly. *See also Premolded Plastic Semiconductor Packages.*

Molded Plug A connector with an electrical cable molded to its end.

Molding Cycle One complete operation of the molding press to produce a molded part. *See also Compression Molding, Injection Molding, Molded.*

Mold Release A lubricant used to coat a mold cavity to prevent the molded plastic piece from sticking to it, and thus to facilitate its removal from the mold. *Also called parting agent, release agent.*

Mold Shrinkage The difference in dimensions, expressed in inches of shrinkage per inch of length, between a molded part and the mold cavity in which it was molded. Both the mold and the part are at room temperature when measured. *See also Mold.*

Molecular Weight The sum of the atomic masses of the elements forming the molecule.

Moly Tab A small, flat piece of gold-plated metal. In hybrid processing, a small die, such as a diode or a transistor, is eutectically bonded to the gold-plated surface of the tab. The moly tab, or Kovar tab, bearing the device is later epoxy-attached to the substrate. Moly tabbing is commonly an operator-assisted

machine operation. *Also called Kovar tab.*

Monofilament A single fiber or filament, as opposed to a braided or twisted filament pair.

Monolithic An electronic packaging technology in which the elements are not readily identified as individual parts.

Monolithic Ceramic Capacitor A multilayer ceramic capacitor.

Monolithic Circuit *See Monolithic Semiconductor Integrated Circuit.*

Monolithic Device A device whose circuitry exists in one die or chip. *See also Multichip Integrated Circuit.*

Monolithic Integrated Circuit (MIC) *See Monolithic Semiconductor Integrated Circuit.*

Monolithic Microwave Integrated Circuit (MMIC) An integrated circuit similar to a monolithic semiconductor integrated circuit, but for use in microwave functions and applications. *See also Monolithic Semiconductor Integrated Circuit.*

Monolithic Semiconductor Integrated Circuit An integrated circuit consisting exclusively of elements formed in situ on or within a single semiconductor substrate with at least one of the elements formed within the substrate. *Also called monolithic integrated circuit.*

Monomer A small molecule that is capable of reacting with similar or other molecules to form large, chainlike molecules called polymers. *See also Plastic, Polymer.*

Mother Board A printed circuit board used for plug-in cards or daughter boards and subsequent interconnecting terminations between them. *See also Backplane Panel.*

Mounting Hole A hole that is used for the mechanical support of a printed board or for the mechanical attachment of components to a printed board.

Multichip A device that contains more than one chip or is capable of performing the functions of several chips. *See also Multichip Module, Multichip Package.*

Multichip Hybrid Package *See Multichip Package.*

Multichip Integrated Circuit An integrated circuit whose elements are formed on several semiconductor chips and individually attached to a substrate or package. *See also Monolithic Device.*

Multichip Microcircuit A microcircuit made up entirely of active and passive chips that are individually attached to the substrate and subsequently interconnected to form the circuit.

Multichip Microelectronic Assembly A package containing several discrete active electronic devices mounted on a substrate and inter-connected with thin- or thick-film circuitry. These devices and circuits are interconnected to the circuitry by thermocompression or ultrasonic bonds. *See also Bond, Thick Film, Thin Film.*

Multichip Module (MCM) A package containing several chips on a ceramic or other type of substrate. Most commonly applied to the use of very large scale integrated (VLSI) circuit chips. MCMs are classified into three categories, based on the type of interconnecting substrate used:

MCM-L: High-density, laminated printed circuit substrate;

MCM-D: silicon substrates or ceramic and metal substrates with deposited interconnect metal circuitry patterns;

MCM-C: Ceramic substrates, either cofired or low-dielectric constant ceramic

Multichip Package An electronic package containing more than one chip. One related term, *multichip hybrid*, or simply *hybrid* defines a package containing a hybrid combination of attached and deposited microcomponents, including chips. These chips have a complexity below that of very large scale integrated (VLSI) circuit chips. A second related term, *multichip module*, defines a package containing more than one VLSI chip. *See Hybrid Circuit, Multichip Module.*

Multifiber In fiber optics, a coherent bundle of fused fibers that function mechanically as a single glass fiber.

Multifunctional Epoxy An epoxy resin having more than two epoxide reactant units per epoxy molecule. This increased number of reactant units per molecule results in a cured epoxy having increased thermal stability over that obtained with difunctional epoxies. *See also Difunctional Epoxy, Trifunctional Epoxy.*

Multilayer *See Multilayer Board.*

Multilayer Board A circuit board or substrate consisting of layers of electrical conductors separated from one another by insulating supports and fabricated into a solid mass. Interlayer connections are used to establish continuity between various conductor patterns. *See also Cofiring, Multilayer Ceramic, Multilayer Printed Circuits.*

Multilayer Ceramic (MLC) A ceramic substrate containing multilayers of thick-film circuitry separated by dielectric layers and interconnected by vias. *See also Cofiring.*

Multilayer Ceramic Capacitor A ceramic capacitor consisting of several thin layers of ceramic. After the layers are assembled and have electrodes positioned, the assembly is fired.

Multilayer Metallization A metallization pattern in which the conductive network is fabricated in more than one plane and separated, except at desired contact points, by thin dielectric films. *See also Metallization.*

Multilayer Printed Circuits Electric circuits made on thin copper-clad laminates, stacked together with intermediate prepreg sheets and bonded with heat and pressure. Subsequent drilling and electroplating through the properly registered layers results in a multilayer circuit board. *See also Registration.*

Multilayer Substrate *See Multilayer Ceramic.*

Multimode In fiber optics, a single fiber that is capable of propagating several modes of a given wavelength.

Multimode Fiber *See Multimode.*

Multiple Circuit Layout A layout consisting of an array of the same circuits on a substrate.

Multiple Conductor Cable Two or more conductors or wires that are insulated from one another and also insulated from the outer covering.

Multiple Termination Module (MTM) A device that, when heated, gang-solders terminations and also insulates and environmentally seals the terminations of a flat conductor cable.

Multiplexer An electrical component that allows two or more coincident signals to be transmitted over a single channel.

Multiplexing The combining of two or more signals from several channels into a single channel for transmission.

129

Munsell Color System A color specification system used in photography, color printing, painting, and other processes. It is based on Munsell hue, value, and chroma when viewed under specific conditions.

Mutual Capacitance The capacitance between two conductors, with all the remaining conductors electrically connected and considered to be grounded.

Mylar A polyester film or tape made by DuPont.

N

Nailhead Bond *See Ball Bond.*

Nano- A prefix meaning one-billionth (1×10^{-9}).

Nanosecond The rise time of a signal or time delay, equal to 10^{-9} or 0.000000001 second.

NC Contacts Normally closed contacts, with the power "on" in the control circuit. *See also NO Contacts.*

N-Channel Charge-Coupled Device A charge-coupled device fabricated so that the charges stored in the potential wells are electrons.

Near Infrared Reflectance Analysis (NIRA) The spectral region having a wavelength of approximately 1 to 3 micrometers in which infrared spectroscopy analysis is performed. Typical infrared spectrophotometers use a light source with a wavelength of approximately 3 to 15 micrometers.

Neck Break Bond A wire bond that fails above the ball bond of a thermocompression bond.

Necking Localized reduction of the cross-sectional area of a tensile specimen during loading.

Negative-Acting Resist A light-sensitive material that is polymerized by a specific wavelength of light and that, after exposure and development, remains on a surface in those areas that were under the transparent areas of a production master. *See also Positive-Acting Resist.*

Negative Image A reverse print of the circuit in which the clear or transparent areas correspond to the circuitry or conductive elements, while the dark or opaque areas correspond to nonconductive base material. *See also Positive Image.*

Negative Temperature Coefficient The condition of a material in which other properties such as the physical dimensions and electrical resistance decrease as the temperature of the material is increased. *See also Positive Temperature Coefficient.*

NEMA Standards Property values adopted as standard by the National Electrical Manufacturers Association (NEMA). One main example is the set of standards for clad and unclad laminates, such as FR-4 epoxy glass, used in printed wiring boards.

Neoprene A synthetic rubber that has good resistance to oil and chemicals and is also flame retardant.

Nest *See Anvil.*

Net Terminals that are interconnected to a common DC power source in an electronic package.

Network A combination of electrical elements such as interconnected resistors and capacitors to form interrelated circuits.

Newtonian Fluid A liquid for which the rate of shear is proportional to the shearing stress. The constant ratio of the shearing stress to the rate of shear is the viscosity of the fluid or liquid.

Next Higher Assembly An assembly of the next higher level in the breakdown of a drawing system.

Nick An undesirable cut or notch in a conductor or its insulation jacket.

Nickel A metallic element with an atomic number of 28. It is corrosion and temperature resistant and is used for alloying purposes, such as iron-nickel alloys widely used in electronic packaging. *See also Invar, Kovar.*

Noble Metal (1) A metal such as gold or platinum that is highly resistant to corrosion and oxidation. (2) Certain high-performance alloys such as stainless steel and certain brass and bronze alloys. *See also Precious Metals.*

Noble Metal Paste Thick-film conductor paste composed of noble metals such as gold and platinum, which are highly resistant to corrosion and oxidation. *See also Thick Film.*

No-Clean Solder Flux *See Flux, No-Clean.*

NO Contacts Normally open contacts, with power "off" in the control circuit. *See also NC Contacts.*

Noise (1) An undesirable, high-frequency disturbance in an electrical system that modifies or affects the performance of a desired signal. (2) Acoustic noise resulting from PIND testing. *See also Particle Impact Noise Detection (PIND).*

Nomex The trademark name for a group of DuPont aromatic polyimides in paper form used in the manufacture of laminated forms. Basic characteristics are similar to Kevlar. *See also Kevlar.*

131

Nominal Resistance Value The specified resistance value of the resistor at its rated load.

Nominal Value The specified value as opposed to the actual value.

Nonactivated Flux A natural or synthetic resin flux without activators. *See also listings under Flux.*

Non-Amine-Cured Epoxy An epoxy resin cured with agents other than amine. Commonly used to describe epoxy laminates cured with phenolic aromatic amino hardeners, which process similar to laminates cured with dicyandiamide hardeners (FR-4 laminates) but exhibit higher thermal stability. *See also Difunctional Epoxy.*

Nonconductive Epoxy An epoxy resin with or without a filler that does not contain electrically conductive fillers such as aluminum powder. A filler is often added to improve thermal conductivity or other properties of resins. *See also Conductive Epoxy, Filled Plastic, Filler.*

Noncontaminating Compound A material that is inert to surrounding materials so as to prevent leaching, contamination, degradation, or other undesirable reactions of the compound under applied environmental conditions. *See also Contaminant.*

Nonfunctional Interfacial Connection A plated through hole in a double-sided printed circuit board that electrically connects a printed conductor on one side of the board to a nonfunctional land on the other side of the board.

Nonfunctional Land A land located on a layer that is not connected to any conductive pattern. *See also Land.*

Noninductive Components Components designed to eliminate the inductive field that they would otherwise create.

Nonionic *See Nonpolar.*

Nonlinear Device A device or circuit in which an increase in applied voltage does not produce a proportional increase in current. *See also Linear Integrated Circuit.*

Nonlinear Dielectric A ceramic material with a nonlinear capacitance-to-voltage characteristic. A barium titanate ceramic capacitor is a nonlinear dielectric. *See also Nonlinear Device.*

Nonpolar Pertaining to a substance that does not ionize, or ionizes very little in water. *See also Nonpolar Solvent, Polar, Polar Solvent.*

Nonpolar Solvent A solvent that is not electrically conductive. It will dissolve nonpolar compounds such as oils, waxes, and resins, but will not dissolve polar compounds. Typical are chlorinated solvents used either as liquids or in vapor degreasers. *See also Polar Solvents.*

132

Nonvolatile Memory A memory in which the data content is retained when power is no longer supplied to it. *See also Volatile Memory.*

Nonwetting Surface A low-surface-energy condition that allows minimum adhesion of materials to the surface. One excellent example is the surface of Teflon, to which solders or adhesives will not adhere. The surface of Teflon must be treated to make it bondable. *See also Low-Surface-Energy Materials.*

Normalize Adjustment of data to a common reference point from which standard intervals of data are applied.

Notch *See Kerf.*

Notch Sensitivity The extent to which the sensitivity of a material to fracture is increased by the presence of defects or discontinuities, a sudden change in section, a crack, or a scratch. Low notch sensitivity is usually associated with ductile materials, and high notch sensitivity with brittle materials.

Novolac Epoxy An epoxy resin having a higher degree of crosslinking, and hence a higher thermal stability, than difunctional bisphenol epoxies. *See also Difunctional Epoxy.*

N-Type Semiconductor Material A crystal of pure semiconductor material to which an impurity, such as arsenic or phosphorous, has been added, thereby changing its characteristics. Electrons are the majority of changed carriers. *See also P-Type Semiconductor Material.*

Nuclear Electromagnetic Pulse (NEMP) The electromagnetic field that is formed from a nuclear blast.

Nuclear Radiation Resistance The ability of a material or part to withstand nuclear radiation and still perform its designated function.

Nugget A small area of recrystallized material at a bond interface.

Numerical Aperture (NA) With respect to an optical conductor, the degree of openness of the input acceptance cone at the end of the fiber or its acceptance of impinging light.

Nylon The generic name for all synthetic polyamides. These thermoplastic polymers have a wide range of good properties, such as abrasion resistance and chemical resistance.

O

Occluded Contaminant A contaminant that has been absorbed by a material. *See also Contaminant.*

O Crimp An O-shaped insulated support crimp.

Off Bond *See Mislocated Bond.*

Off Contact In screen printing, the preset space or gap between the screen and substrate. *Also called breakaway.*

Ohm The unit of electrical resistance. The resistance through which one ampere of current will flow when a voltage of one volt is applied.

Ohmic Contact An electrical connection between two materials across which the voltage drop is the same in either direction.

Ohms per Square The unit of measurement for sheet resistivity or surface resistivity. It is the resistance of a square area, measured between parallel sides of both thin- and thick-film resistive materials. *See also Sheet Resistivity, Surface Resistivity.*

Olefin A family of unsaturated hydrocarbons with the formula C_nH_n, named after corresponding paraffins by adding *-ylene* or *-ene* to the stem (e.g., ethylene). Paraffins are aliphatic hydrocarbons. *See also Hydrocarbon.*

Oligomer A polymer containing only a few monomer units, such as a dimer or trimer. *See also Monomer, Polymer.*

Olyphant Washer Test A simple and practical test developed by Murray Olyphant of 3M Company for testing the comparative crack resistance of casting and potting resins. Four V-shaped notches are machined in steel test washers, which providing sharp edges that cause crack-prone resins to crack during thermal cycling or thermal shock. It is simpler and less costly than the Hex Bar Test. *See also Hex Bar Test.*

Onsertion The positioning of a surface-mount component onto the surface of a printed wiring board or substrate with respect to the solder lands. Accuracy in onsertion is critical.

Opacity The ability of a material to resist the entrance or penetration of light waves.

Open *See Open Circuit.*

Open Barrel Terminal A terminal with an open insulated barrel to accommodate a wire and to be subsequently crimped to form a reliable electrical connection.

Open-Cell Material A foamed or cellular material made up of cells that are interconnected. *See also Closed-Cell Material.*

Open Circuit A circuit that has been disrupted so that no current flow exists. *See also Short Circuit.*

Open-Entry Contact A female-type contact that is unprotected from damage caused by probes and other devices.

Operator Certification A program whereby operators are trained and certified to perform specific functions, such as wire bonding, chip mounting, soldering, and thick-film screen printing.

Opposed-Electrode Welding A resistance weld in which the parts to be welded are placed between two opposed electrodes. The current is passed between the electrodes and through the parts, which are heated and thereby welded from the resistance of the interfaces of the parts.

Optical Comparator A microscopic device that is capable of projecting a magnified image of the workpiece on a screen so that dimensions, surface flows, and other properties can easily be measured.

Optical Conductors In fiber optics, materials that have a low optical attenuation to the transmission of light energy.

Optical Coupling In fiber optics, the leakage of light from one fiber into another. *Also called crosstalk.*

Optical Interconnects Optoelectronic devices such as a light-emitting diode (LED) and an optical coupler. The LED converts an electrical signal into light at the electrooptical interface, while the optical coupler couples signals from one electronic circuit to another through the use of an LED plus a phototransistor. This combination converts an electrical signal of the primary circuit into light, and the phototransistor in the secondary circuit reconverts the light signal into an electrical signal.

Optical Mosaic In fiber optics, a construction of fibers into a group or groups, with some degree of imperfection occurring at the boundaries of the groups. When imperfections reach a high level, the construction is called *whichen wire.*

Optoelectronic Pertaining to devices containing both optical and electronic components resulting from electrooptic technology. Two of the many examples are light-emitting diodes and electroluminescent displays. *See also Electrooptics.*

Orange Peel (1) A rippled surface texture of solder mask material over circuits containing tin or tin-lead plating on printed wiring boards. The orange peel indicates the reflow of the plated metals under the solder mask material. (2) An undesirably rough surface on a molded part, resembling the surface of an orange.

Organic (1) Composed of matter originating in plant or animal life, or composed of chemicals of hydrocarbon origin, either natural or synthetic. Used in referring to chemical structures based on the carbon atom. (2) A chemical structure built upon the carbon atom. Most polymers are organic. Silicones, as an example of partially inorganic polymers, have a chemical structure that is built around the silicon atom. Organics burn to a black carbon ash, while silicones burn to a whitish silicon-dioxide ash. *See also Inorganic Chemicals.*

Organic Composites The family of fiber structures that include continuous or discontinuous fibers in an organic resin matrix material, such as graphite-epoxy composites. Applications include commercial and military products that must be lightweight, strong, and rigid. *Sometimes called advanced composites.* Printed wiring board laminates are simple composites. *See also Composites, Laminated Plastics, Organic.*

Organic Flux A flux composed of a rosin base and a solvent. *See Flux, Organic Acid.*

Organic Pigments Dyes or coloring materials with excellent color stability, used to color plastics. *See also Inorganic Pigments.*

Organic Vehicle (1) The rosin base material of a solder flux. (2) The solvent carrier of a paint.

Organic Water-Soluble Flux *See Flux, Organic Acid.*

O Ring A rubber ring used in a very broad range of applications that require an O-shaped seal, as in piping and tubing.

Outer Lead Bonding (OLB) In tape-automated bonding, the bond that attaches the lead to the substrate. The inner lead bond (ILB) is the bond that attaches the lead to the chip. *See also Inner Lead Bond, Tape-Automated Bonding.*

Outgas The release of gases and vapors from a material over a period of time when subjected to temperatures above approximately 100°C, to vacuum conditions, or to both. Outgassing can be forced or natural, such as occurs in space-borne systems.

Ovaled Pertaining to an elliptically shaped terminal or contact designed to accommodate two wires.

Oven Soldering A process in which multiple solder terminations are made simultaneously. Use of the process is limited, however, since few materials and components can withstand the high temperatures required for the length of time to which they would be exposed.

Overbonding *See Chopped Bonds.*

Overcoat A thin film of dielectric material that is applied over circuitry and components for environmental, mechanical, and contamination protection. *See also Conformal Coating.*

Overcurrent An excessive amount of current in a conductor, causing a rise in temperature of the conductor and its insulation.

Overetching *See Undercut.*

Overflow Wire A wire connection that is called out on the wiring diagram but not actually connected during automatic wiring of the package.

Overglaze A printed and fired glass layer that is applied over resistors or used as a solder barrier in thick-film technology. *See also Thick Film.* See Fig. 12.

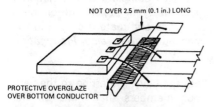

NOT OVER 2.5 mm (0.1 in.) LONG

PROTECTIVE OVERGLAZE
OVER BOTTOM CONDUCTOR

Figure 12: Overglaze

Overlap The overlay or contact area between the thick-film resistor and conductor. *See also Thick Film.*

Overlay The application of one material over another material.

Overpotential *See Overvoltage.*

Overspray (1) The undesired spreading of the abrasive from the nozzle of a resistor-trimming machine onto adjacent resistors. (2) Undesirable excess coating materials resulting

from poorly controlled spray-coating condition.

Overtravel (1) In screen printing, the excess distance a squeegee travels in the *Y* (lateral) direction, beyond the pattern on the substrate, before the squeegee lifts off the screen. (2) The excess distance in the *Z* (downward) direction that a squeegee blade would push the screen if the substrate and nest plate were not in place.

Overvoltage A voltage greater than normal operating voltage. *Also called overpotential.*

Oxidation A process in which a metal reacts with oxygen in the atmosphere to form an oxide. For example, iron reacts with oxygen to form iron oxide or rust. *See also Reducing Atmosphere.*

Oxidizing Atmosphere An air- or oxygen-enriched atmosphere in a furnace.

Oxygen-Free High-Conductivity (OFHC) Copper Oxygen-free high-conductivity copper having a minimum copper content of 99.95 percent. Normally used in electronic applications.

Ozone (1) A reactive form of oxygen that is produced by an electrical discharge. (2) A layer of the upper atmosphere that limits the amount of solar radiation to the earth.

Ozone-Depleting Chemicals Those chemicals that destroy the atmo-

spheric ozone layer, thereby allowing excess solar radiation to reach
the earth.   Chlorofluorocarbon
(CFC) chemicals used in solder
cleaning are among the worst of-
fenders. *See also Chlorofluorocarbons.*

Ozone Test A test in which materials are exposed to high concentra-
tions of ozone, which produces
accelerated indication of degradation.

P

Package (1) An enclosure for electronic components and hybrid circuits consisting of a header, a lid,
and hermetically sealed feedthrough
terminal leads. Packages are made
of metal, ceramic, and plastic. (2)
An enclosure or housing used to
contain any level of electronic system or subsystem. *See also Package Level.*

Package Cap *See Lid.*

**Package and Interconnecting Struc-
ture** The generic term for a completely processed combination of
substrates, metal planes, or constraining cores and interconnection
wiring used for the purpose of
mounting components. *See also
Constraining Core.*

Package Crossing   An electrical
connection from a terminal on one
package to a terminal on another
package.

Package Delay   The time delays
caused by the interconnections and
distance between components in
order to complete their functions.
This delay is dependent on both
distance and materials used in the
package. *See also Propagation
Delay.*

Package Level Referring to the various members that make up the
packaging hierarchy—such as the
chip, chip carrier, printed wiring
board, mother board, chassis, and
system in order from a low level to
a high level.

Package Lid *See Lid.*

Packaging The process of physically
locating, connecting, and protecting
electronic components. *See also
Electronic Packaging.*

**Packaging and Interconnecting (P/I)
Assembly** An assembly that has
components mounted on one or
both sides of an interconnecting

structure, such as a printed wiring board.

Packaging and Interconnecting (P/I) Structure A printed wiring board or similar base of interconnecting circuitry.

Packaging Density The quantity of components, interconnections, and mechanical devices per unit volume. Classified as high, medium, or low density.

Pad (1) A widened area of a conductor on a substrate used for attaching wire bonds, mounting and bonding chip devices, or contacting with test probes. (2) A portion of the conductive pattern of printed circuits for mounting components. *Also called land.*

Pad Grid Array (PGA) A pinless package whose electrical contact to the substrate is made with screened solder contact pads over the entire bottom in a checkerboard array. Each pad is located where a via exits the ceramic or other substrate. Used in the highest input/-output (I/O) devices and packages. *See also Pin Grid Array.*

Panel (1) A metal plate on which connectors are mounted. (2) The base material from which one or more printed wiring board assemblies are made after the metallized laminate has gone through the etching or circuit-forming process.

Panel Mount The attaching of the female half of a connector to a panel.

Panel Plating The plating of the entire surface of a panel, including holes. *See also Panel, Pattern Plating.*

Parallel-Gap Solder A method in which excessive current is passed through a high-resistance gap between two electrodes, the heat from which reflows the solder and forms an electrical connection.

Parallel-Gap Weld A weld formed by passing excessive current through a high-resistance gap between two spring-loaded electrodes, which in turn applies a mechanical force to the component leads or conductors, causing them to weld together into an electrical connection. *See also Percussive Arc Welding.*

Parallelism The amount of variation in thickness of a substrate, wafer, or other component.

Parallel Splice A splice in which a holding tool is used to ensure that the conductors lie parallel to each other during joining.

Parameter An operating condition or characteristic with various values that influence operating performance.

Parasitic Losses Electrical losses in a circuit caused by undesirable features and characteristics of the packaging, interconnection, and materials used. Examples of these causes are high-dielectric-constant materials, long lead lengths, and high-resistance interconnection joints. *See also Dielectric Constant.*

Part *See Component.*

Partial Discharge A high-voltage electrical discharge that partially bridges the insulation between conductors. *Also called corona.*

Partial Discharge Pulse A voltage or current pulse as a result of a partial discharge.

Partial Lift A wire bond partially raised from the bonded pad.

Particle Impact Noise Detection (PIND) The detection of undesirable loose particles in a sealed microcircuit package by sensitive acoustic testing. Typical particles are dislodged wire or conductor fragments.

Parting Agent In the molding of a plastic part, a lubricant that is applied to the surface of the mold to prevent the finished parts from adhering to the mold. *Also called mold release or release agent.*

Parting Line In the molding of a plastic part, the line formed in the part by the mating surfaces of the mold halves.

Partitioned Mold Cooling The cooling of the mold by circulating water through the core.

Parts Density The number of parts per unit of volume. *See also Packaging Density.*

Parylene A polymer resin, polyparaxylene, which provides a very thin (250–500 angstrom), uniform conformal coating on printed wiring assemblies and electronic components. It is applied by vacuum deposition and provides a very uniform coating on sharp edges, on complex shapes, and in holes. This beneficial feature is unique among circuit board coating processes. Trade-named by Union Carbide Company.

Passivated Region Any region covered by glass, silicon dioxide, nitrides, or other protective insulating coating materials.

Passivation (1) The application of an insulating layer of glass, SiO_2, or nitride over circuits and circuit elements for protection against moisture, contaminants, or other harmful conditions. (2) The growth of an oxide or nitride layer on the surface of a semiconductor to provide electrical stability by isolating the surface from electrical and chemical conditions in the environment. *See also Glassivation.*

Passive Components Electrical components that do not change their character when an electrical impulse is applied. Examples are resistors, capacitors, and inductors. *See also Active Components.*

Passive Network (1) A network that has no source of energy. (2) A network of passive elements, such as screened resistors, that are interconnected by conductors.

140

Passive Substrate A substrate that acts as a support of the circuitry or as a heat sink and does not have any active devices. Usually made of alumina, ceramic, or glass. *See also Substrate.*

Passive Trimming Adjusting the function of a circuit while the power is turned off.

Paste A thick-film screen-printable composition of micron-size polycrystalline solids suspended in a thixotropic vehicle. *Often referred to as ink or composition.* The paste may contain metals, metal oxides, or glasses, depending on whether its use is for deposited conductors, resistors, or dielectrics. *See also Cermet Thick Film, Polymer Thick Film, Thick Film.*

Paste Blending *See Blending.*

Paste Soldering A process in which a paste, composed of solder particles in a flux, is screen-printed onto a film circuit and reflowed to form soldered connections to chip components. *See also Reflow Soldering, Solder Paste.*

Paste Transfer To pass a thick-film ink or composition through a screen or mask and deposit it in a pattern on a substrate. *See also Thick Film.*

Path A part of a printed circuit between two pads or between a pad and a terminal area.

Pattern The configuration of conductive and nonconductive materials on a substrate.

Pattern Plating The selective plating of a conductive pattern on a substrate. *See also Panel Plating.*

P-Channel Charge-Coupled Device A charge-coupled device fabricated so that the charges stored in the potential wells are holes.

Peak Firing Temperature The maximum temperature in the firing profile of thick-film pastes. *See also Thick Film.*

Pebble Mill *See Roller Mill.*

Peel Imposition of a tensile stress in a direction perpendicular to the adhesive bond line of an etched printed wiring board, a circuit pattern, or a flexible adherend in an adhesive bonded structure. *See also Adherend, Adhesive.*

Peel Strength A measure of the force required to separate two adhesive bond materials, as determined by pulling or peeling the materials. The unit of measure is oz/mil or lb/inch of width.

Penetration The entering of one part or material into another, as in certain hardness tests. *See also Indentation Hardness.*

Percent Defective Allowable (PDA) The maximum percentage of defective parts that will permit the lot to be accepted after the specified 100 percent test.

141

Percussive Arc Welding A process in which a fixed gap is maintained between the surfaces of two parts to be welded while radio frequency energy is applied. *See also Parallel-Gap Weld.*

Perforated Terminal A flat metal solder terminal with an opening through which one or more wires are placed prior to soldering.

Perimeter Sealing Area The sealing surface on an electronic package. It is located along the perimeter of the package cavity and defines the area to which the lid or cover is bonded or welded. *See also Electronic Package.*

Periodic Table A listing of chemical elements, showing their atomic number and atomic weight and indicating their electron structure.

Peripheral Solder Sealing A flux-free, normally manually operated process, applied as a rule to plat-form-type package sealing. The operator uses solder-core wire inside a glove box while the oxygen contact inside the box is limited.

Permanence The resistance of a given property to deteriorating influences.

Permeability (1) The property of a material that allows the diffusion or passing through of a vapor, liquid, or solid without physically or chemically affecting the material. (2) The degree to which a metal modifies the magnetic flux in the region

occupied by it in a magnetic field. *Also called magnetic permeability.*

Permittivity *See Dielectric Constant.*

Personal Ground Strap *See Wrist Strap.*

pH A measure of the acidity or alkalinity of a substance, neutral being a pH of 7. Acids range from 0 to 7, while alkaline or base solutions range from 7 to 14. A pH of 1 is most acid and a pH of 14 is most alkaline.

Phase-Change Materials Materials that absorb heat as they change phase, such as from a solid to a liquid. These materials are useful in thermal management of electronic packages and systems.

Phased Array An antenna that forms a beam by assigning phases to a number of separate radiating elements.

Phase Diagram A graphic representation of the compositions, temperatures, and pressures at which the heterogeneous equilibria of an alloy system occur.

Phase Shifter An electronic circuit that shifts phase in specific predetermined steps.

Phenolic (1) A synthetic resin produced by the condensation of an aromatic alcohol with an aldehyde, particularly of phenol with formaldehyde. (2) One of the largest volume-produced thermosets. Extensively used because of its low cost and good electrical insulation characteristics.

Phenylsilane A thermosetting copolymer of silicone and phenolic resin; furnished in solution form.

Phosphor Bronze An alloy of copper, tin, and phosphorus.

Phosphosilicate Glass (PSG) A phosphorous doped silicon dioxide that is sometimes used as a dielectric layer because it prevents the diffusion of sodium impurities. Since it softens and flows at approximately 1050°C, it creates a smooth finish for subsequent layering.

Photo Etch A process in which circuit patterns are formed by exposing (polymerizing, light-hardening) a photosensitive polymer photoresist through a photo positive or negative of the circuit and etching away the part of the film that was not protected by the polymerized material. *See also Photopolymer.*

Photofabrication The fabrication of fine circuit features on electronic devices and substrates through photolithographic techniques. *See also Photolithography.*

Photolithography A technology employed to create a pattern that includes rubylith, photoreduction, step and repeat, computer-aided design, and electron beam techniques. The process produces a mask that can be used to image a microelectronic device.

Photomask *See Mask.*

Photon A fundamental particle of mass zero that is regarded as the quantum of radiant energy. The energy E of a photon is hv, where h is Planck's constant and v is the optical frequency. *See also Fiber Optics, Photonics, Planck's Constant.*

Photonics The branch of science and technology devoted to the study control and use of photons or light waves, including the generation, transmission, reception, and processing of optical signals and power. It includes such technologies as modulation, amplification, and image processing, storage, and detection. *See also Fiber Optics, Photon.*

Photopolymer A polymer that changes characteristics and can be cured or polymerized when exposed to light of a given frequency.

Photoresist Negative *See Negative-Acting Resist.*

Photoresist Positive *See Positive-Acting Resist.*

Phototools Masks and other hardware items used to produce fine features on electronic devices and substrates through the use of photolithography techniques. *See Photolithography.*

Physical Design The location and orientation of chips, packages, devices, and their respective interconnections in an electronic package or in a total system.

Physical Vapor Deposition A vapor deposition process using a mask or collimator between the sputtering target and the wafer, thus resulting in higher-definition deposits for more complex integrated circuits. *See also Chemical Vapor Deposition.*

Pick The distance between two adjacent crossover points of braid wires. The measurement in picks per millimeter (picks per inch) indicates the degree of coverage.

Pick and Place A manufacturing operation in which chips are correctly selected, placed, and oriented on their respective pads on the substrate prior to bonding.

Pickup Interference from a nearby circuit or system.

Pico- A prefix meaning one-trillionth (1×10^{12}).

Pigment A finely divided powdered substance used for coloring and insoluble in the vehicle in which it is used. *See also Inorganic Pigment, Organic Pigment.*

Pigtail (1) The excess wire that extends beyond the bonding pad after a bond is made. (2) A short length of wire used as a jumper or ground, or for terminating purposes.

Pin A small-diameter metal rod used as an electrical terminal and/or a

mechanical support. Pins are used inside a package to support a wire bond and externally as a plug-in type of connection. They may be straight, modified as a nailhead, upset-pierced, or a formed variety.

Pin Contact A male-type contact designed to mate with a female contact.

Pin Density The number of pins on a printed wiring board per unit area.

Pin Grid A two-dimensional arrangement of electrically conductive pins equally spaced and parallel to one another.

Pin Grid Array (PGA) (1) A predetermined configuration of many plug-in electrical terminals for an electronic package or interconnection application. (2) A package with pins located over nearly all of its surface area. (3) A square-format package with leads distributed over the bottom of a grid pattern at 0.1 inch (2.54 mm) or finer pitch. *See also Pad Grid Array, Pin Grid.* See Fig. 13.

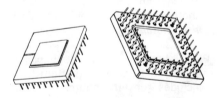

Figure 13: Pin Grid Array

Pinhole (1) A small hole that extends through a printed element to the base material. It can be in both metallized and dielectric materials. (2) A small hole occurring in a semiconductor device as an imperfection that penetrates entirely through film elements such as metallization films or oxide layers. *See also Pits.*

Pitch (1) The nominal distance from center to center of adjacent conductors. (2) The distance between a point on an image and the corresponding point on the corresponding image in an adjacent pattern lying in either a row or a column. *Also called center-to-center spacing.*

Pitch Diameter The diameter of a circle passing through the center of the conductors within the same layer of a multiconductor cable.

Pitch Fibers Carbon fibers made from the high-molecule-weight residue of the destructive distillation of coal or petroleum products.

Pits Small holes that do not extend through printed elements to the base material. *See also Pinhole.*

Placement The placement and correct orientation, whether manual or automatic, of integrated circuits, packages, and circuit cards in their respective locations at a given package level.

Plain Weave The simplest and most commonly used fiber weave in which the warp and filling threads cross alternately. Plain woven fabrics are generally the least pliable, but they are also the most stable. Stability permits the fabrics to be woven with a fair degree of porosity but without too much sleaziness. *See also Basket Weave, Crow Foot Weave, Fiber, Fill, Warp.*

Planar (1) Existing or lying in a single plane. (2) Pertaining to a type of semiconductor device, and the process technology used to fabricate it, in which all P-N junctions terminate at approximately the same geometric plane on the surface of the semiconductors. Devices using similar technologies, but having one or more diffused areas lying in a slightly different, but parallel plane, are also considered planar (e.g., buried collector).

Planar Motor A motor having a flat, planer configuration.

Planar Motor Voice Coil Servo A mechanical positioning device with very high accuracy, good feedback, limited excursion, and very high speed.

Planck's Constant The number h that relates the energy E of a photon with the frequency v of the associated wave through the relation $E = hv \times h = 6.626 \times 10^{-34}$ joule seconds. *See also Fiber Optics, Photon, Photonics.*

Plasma An electrically conductive gas composed of high-energy neutral particles, ionized particles, and free electrons, used to dry-etch or plasma-etch dielectric materials. *See also Plasma Etching.*

Plasma Cleaning *See Plasma Etching.*

Plasma Erosion Uncontrolled erosion of materials resulting from the high energy of ionized gas.

Plasma Etching A dry, controlled etching process in which plastic substrates, such as epoxy-glass printed wiring boards, are exposed to ion bombardment by a gas in a vacuum to improve bondability during electroplating. *See also Etching.*

Plastic A polymer blended with all of the additives required for a final product. Additives may include plasticizers, flame retardants, fillers, and colorants. *See also Polymer, Thermoplastic, Thermoset.*

Plastic Deformation A change in dimensions of an object under load that is not recovered when the load is removed, as opposed to elastic deformation. *See also Elastic Deformation.*

Plastic Device A molded electronic package made of plastics, such as epoxies, phenolics, and silicones, that contains semiconductor or electronic components. Examples are plastic leaded chip carrier (PLCC) and plastic quad flat pack (PQFP).

Plastic Encapsulation The embedding of an electronic assembly in a plastic material for ruggedization and environmental protection.

Plasticity A property that allows a material to be deformed continuously and permanently without rupture upon the application of a force that exceeds the yield value of the material.

Plasticize To soften a material and make it plastic or moldable, either by the addition a plasticizer or the application of heat. *See also Plasticizer.*

Plasticizer A material incorporated in a resin formulation to increase its flexibility or workability. The addition of a plasticizer may cause a reduction in melt viscosity, lower the temperature of second-order transition, or lower the elastic modulus of the solidified resin. *See also Flexibilizer.*

Plastic Leaded Chip Carrier (PLCC) An electronic package, made of plastic, containing a chip that has leads extending from the sides of the package in a typical J-lead or gull wing configuration for surface mounting on printed boards. *See also J Lead, Gull Wing, Lead.* See Fig. 14.

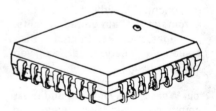

Figure 14: Plastic Leaded Chip Carrier

Plastic Quad Flat Package (PQFP)
As standardized by JEDEC, a package featuring a square body with gull wing leads on all four sides, bumpers on the corners, a slightly thicker body than the quad flat pack, and leads on only 0.025-inch centers. *See also Gull Wing Lead, Quad Flat Pack.*

Plastic Range (1) A temperature range in which most metals can be worked or deformed without causing cracking. (2) A temperature range in which a material is partially liquid and partially solid.

Plastic Shell A thin plastic container used to hold an electronic assembly during subsequent encapsulation. Also used as a container to provide environmental protection but not necessarily to encapsulate the assembly. *See also Plastic Encapsulation.*

Plastisols Mixtures of vinyl resins and plasticizers that can be molded, cast, or converted to continuous films by the application of heat. If the mixtures contain volatile thinners, they are also known as *organosols*. *See also Plasticizer.*

Plated *See Plating.*

Plated Through Hole (PTH) A hole in a multilayer printed board in which an electrical connection is made between internal and external conductive patterns, or both, by the deposition of metal on the wall of the hole.

Platens Flat mounting plates of a plastic molding press that can be adapted to supply heat or cooling and that apply a uniform temperature and pressure to the laminate or molded part.

Plating The process of chemically or electrochemically depositing metal on a surface. Copper, nickel, chromium, zinc, brass, cadmium, silver, tin, and gold are the most common plated or electrodeposited metals.

Plating Anode *See Anode.*

Plating Thief A racking device used in electroplating to provide additional surface area, which in turn will produce more uniform current density on the part to be plated.

Plating-Up A process in which the metal deposition is increased in thickness by electrolytic plating after the base material has been metallized with a thin conductive layer. *See also Electrolytic Plating.*

Plating Void A small hole or cavity in the plated surface caused by a metal inclusion or a protruding particle in the base material or by contamination of the plating bath.

Platinum A precious, white heavy metal that is thermally stable at high temperatures and provides low, consistent surface resistance, making it ideal for contact applications. It is resistant to corrosion and film formation and therefore can be used to replace gold-plated metal parts.

147

Plenum Ductwork or airspace over a suspended ceiling used for the air return path of an air-conditioning system.

Plug-In The device or subassembly level at which replacement can be made by simply inserting one unit in the place of another. No soldering, welding, crimping, or other fastening means are required.

Plug-In Package An electronic package that can be plugged into or removed from a socket, printed wiring board, or other suitable connector. *See also Electronic Package.*

Plunger Molding *See Transfer Molding.*

Ply The number of individual strands or filaments that are twisted together to form a single thread.

Point-to-Point Panel Wiring (1) A mechanized technique in which the wiring is run in a direct path from one terminal to another without dressing the wire. (2) A method of interconnecting components by routing wires between the connecting points.

Poisson's Ratio The absolute value of the ratio of transverse strain to axial strain resulting from a uniformly applied axial stress below the proportional limit of the material.

Polar Pertaining to a substance that readily dissolves and ionizes in water. *See also Nonpolar, Nonpolar Solvent, Polar Solvent.*

Polarity An electrical condition by which the direction of the current flow can be determined in a circuit.

Polarizing Key *See Polarizing Pin.*

Polarizing Pin A pin that is precisely located on one half of a two-piece connector; the matching hole is precisely located on the other half of the connector. The arrangement ensures the correct assembly of the two halves. *Also called polarizing key.*

Polarizing Slot A notch or slot that is machined into the edge of a printed board, at an exact location, to ensure accurate location and insertion of the mating connector.

Polar Solvent (1) A solvent that can dissolve polar compounds such as inorganic salts. It contains hydroxyl or carbonyl groups and has a strong polarity. It cannot dissolve nonpolar compounds such as hydrocarbons and resins. (2) A solvent that is composed of electrically charged particles called ions and that is capable of conducting electricity. Water is a typical example, since it will dissolve acidic and basic materials, resulting in an ionic solution that can be used for electroplating and other processes. *See also Nonpolar Solvent.*

Poly- A prefix identifying most thermoplastic polymers. *See also Polymer.*

Polyacrylonitrile (PAN) A synthetic fiber used as a base material or precursor in the manufacture of certain carbon fibers for advanced composites. *See also Composite.*

Polyamide The chemical or generic name for Nylon. *See also Nylon.*

Polyamide-imide A plastic with outstanding thermal stability at high temperatures and good electrical properties. It has an aromatic structure that when heat-cured, forms a linear amide-imide homopolymer. *See also Aromatic.*

Polyarylsulfone A thermoplastic resin with good chemical, solvent, and impact resistance. It is composed of aromatic phenyl and biphenyl groups, which are linked together by thermally stable ether and sulfone groupings. *See also Aromatic.*

Polybutadiene A thermosetting resin that can be molded by transfer or compression. These resins have excellent electrical properties, good mechanical properties, and outstanding resistance to water and other liquids. They also are noted for their heat resistance of up to 500°F and their temperature stability.

Polycarbonate Resin A thermoplastic resin with high strength and dimensional stability over a wide range of temperatures and humidity. Because of their outstanding electrical properties, heat stability, impact strength, and creep resistance, these resins are excellent candi-

dates for many electronic applications.

Polycrystalline A material composed of a large number of crystals. Alumina, ceramic, and semiconductor materials are examples.

Polyesters Thermosetting resins, produced by reacting unsaturated, generally linear hydrocarbon, alkyd resins with a vinyl-type active monomer such as styrene, methyl styrene, or diallyl phthalate. Cure is effected through vinyl polymerization using peroxide catalysts and promoters, or heat, to accelerate the reaction. The resins are usually furnished in liquid form. *See also Hydrocarbon, Linear.*

Polyethylene A thermoplastic material with excellent resistance to chemicals and moisture, flexibility at low temperature, high electrical resistivity, good dielectric properties at high frequencies, and relatively low cost. These characteristics make polyethylene an excellent candidate for high-frequency electronic applications.

Polyimide A high-temperature thermoplastic resin made by reacting aromatic dianhydrides with aromatic diamines. It is used with glass fibers in the manufacture of printed circuit laminates and provides excellent resistance to wear and oxidation, high temperature stability, weathering resistance, and a low dielectric constant. These properties reduce propagation delay in multilayer construction. *See also*

149

Aromatic, Dielectric Constant, Propagation Delay.

Polymer A compound formed by the reaction of simple molecules having functional groups that permit their combination to proceed to high molecular weights under suitable conditions. Polymers may be formed by polymerization (*addition polymer*) or polycondensation (*condensation polymer*). When two or more monomers are reacted, the product is called a *copolymer*. Polymers can be plastics, elastomers, or liquids. Their chemical structures may be linear or aromatic. *See also Aromatic, Elastomer, Linear, Plastic, Thermoplastic, Thermoset.*

Polymerization Shrinkage Volumetric shrinkage of plastic parts during the polymerization of base materials, caused by shortening of chemical bonds during the polymerization or curing reaction.

Polymerize To unite chemically two or more monomers or polymers of the same kind in order to form a molecule with higher molecular weight.

Polymer Reversion The irreversible softening or liquefaction of a polymer as a result of hydrolysis when the polymer is bombarded with vapor molecules that contain an active hydroxyl group. Hydrolytic stability is required in polymers to overcome this undesirable reversion. *See also Hydrolytic Stability, Hydrophilic, Hydrophobic, Reversion.*

Polymer Thick Film (PTF) A thick-film deposition formed by curing polymer-based inks or pastes onto suitable substrates. Much lower curing temperatures (under 150 to 200°C) are involved than with cermet thick films, which require firing at 700 to 800°C. *See also Cermet Thick Film.*

Polymethyl Methacrylate A transparent thermoplastic composed of polymers of methyl methacrylate. *Sometimes also called acrylic.*

Polynary Pertaining to materials that have many elements, as opposed to binary materials, which have two elements, or ternary materials, which have three.

Polyphenylene Oxide A thermoplastic resin that has excellent electrical characteristics and dimensional stability from −275 to +375°F. It is especially suitable for high temperature and high electrical frequency applications.

Polyphenylene Oxide (PPO) Based Resins A group of thermoplastics having a very low specific gravity. They are tough, rigid materials with excellent mechanical and dimensional stability up to 300°F and low creep and moisture absorption. Electrical properties include high dielectric strength, low dissipation factor, and low dielectric constant up to 1 mHz.

Polyphenylene Sulfide Resin An aromatic polymer with excellent high-temperature and chemical resistance, service temperature to

450°F and melting temperature of 550°F. It can be filled with glass-reinforced fibers for electrical applications.

Polypropylene A plastic that is made by the polymerization of high-purity propylene gas in the presence of an organometallic catalyst at relative low pressures and temperatures. It is noted for its low electrical losses, high tensile strength, and resistance to abrasion, moisture, and heat.

Polystyrene A thermoplastic produced by the polymerization of styrene. It has excellent electrical properties, good dimensional stability, and moisture resistance.

Polysulfone A thermoplastic resin that is flame resistant, is heat resistant to over 300°F for extended periods of time, and has good dimensional stability at elevated temperatures.

Polyurethane Resins A family of resins used to form thermosetting materials by reacting them with water, glycols, or other urethanes. Widely used as elastomers or low-density foams.

Polyvinyl Chloride (PVC) A thermoplastic material composed of polymers of vinyl chloride. It can be blended with other polymers to impart abrasion resistance, heat stability, low shrinkage, and moisture resistance. PVC can be converted into a colorless sheet or film by heat and pressure.

Porcelain A glassy, vitreous, ceramic material. Many varieties are available, such as alumina porcelain, borosilicate, cordierite porcelain, forsterite porcelain, steatite porcelain, titania porcelain, and zircon porcelain.

Porcelain Enamel Technology The technology of coating and bonding a vitreous, glassy, inorganic porcelain material to metal by fusion at temperatures above 800°F. *See also Porcelain.*

Porcelainized Steel Substrates Steel plates or substrates coated with porcelain. These dimensionally stable substrates are subsequently metallized to form stable deposited or etched circuits. *See also Porcelain.*

Porosity (1) Multiple voids in a material. (2) Quantitatively, the ratio of solid matter to voids in a material.

Positive-Acting Resist A light-sensitive material that decomposes or softens by a specific wavelength of light and that, after exposure and development, is removed from those areas of a surface that were under the treatment of a production master. *See also Negative-Acting Resist.*

Positive Image The true or exact picture of a circuit pattern. *See also Negative Image.*

Positive Temperature Coefficient The condition of a material in which other properties such as the physical dimensions and electrical resistance increase as the temperature of the material is increased. *See also Negative Temperature Coefficient.*

Positive Temperature Coefficient Resistor A type of resistor whose resistance increases as the temperature of the resistor increases.

Post *See Terminal.*

Postcuring The additional curing at the cure temperature or elevated temperatures to fully cure and achieve maximum properties of a plastic.

Postfiring The additional firing of thick-film circuits, usually to change the values of resistors.

Poststress Electrical Aging The application of electrical power to a film circuit in order to stress the resistors and stabilize their characteristics or cause failure to marginal devices. *See also listings under Burn-In.*

Post-Type Terminal A terminal to which, after the wire has been wrapped around it, a threaded nut or similar device is added to secure the wire to the terminal.

Pot (1) To embed a component or assembly in a liquid resin, using a shell, can, or case that remains an integral part of the product after the resin is cured. (2) The sealing of the end of a cable connector, having multiple contacts, with a plastic compound to exclude mois-

ture, prevent short circuits, and provide mechanical support to the cable. *See also Cast, Embed.*

Potential The difference in voltage between two points in a circuit when one of the two points is at a zero potential.

Pot Life The time during which a liquid resin remains workable as a liquid after catalysts, curing agents, promoters, or other elements are added; roughly equivalent to gel time. *Also called working life. See also Gel Time.*

Potting Compound Materials used for the embedment or encapsulation of components and wires. They are also used for mechanical support, prevention of electrical shorts, thermal management, and sealing against moisture. These materials are usually applied as liquids, which are then connected to solids by a polymerization or curing reaction. *See also Embed, Encapsulate.*

Potting Cup (1) A metal or plastic cup used to house electronic assemblies that are to be potted. (2) A form attached to the rear or back of a receptacle that provides a moldlike cavity for potting the wires and wire entry of the assembly. *See also Pot.*

Powder Pressing *See Dry Pressing.*

Power Cycling An accelerated reliability testing method in which microelectronic components are placed under cyclic stress by applying electrical power, on an intermittent basis, to a specific heat-generating component in that assembly. *See also listings under Burn-In, High Temperature Reverse Bias Test.*

Power Density (1) The power generated per unit volume in an electronic package. (2) In thick-film thermal management, the amount of power dissipated from a film resistor through the substrate. It is measured in watts per square inch. *See also Electronic Package, Thick Film.*

Power Dissipation The dispersion of heat generated in a device, component, or circuit whose current flows through it.

Power Distribution Those conductors within a package that carry the electrical power to the circuits.

Power Factor (1) The ratio of resistance to impedance. (2) The ratio of the actual power of an alternating current to the apparent power. (3) Mathematically, the cosine of the angle between the voltage applied and the resulting current. This is numerically equivalent to dissipation factor for most insulating materials. *See also Dissipation Factor.*

Power Hybrid Package (PHP) A geometrically shaped metal (usually copper) electronic package containing transistors, resistors, or other discrete components that are soldered directly to the base of the package or to a beryllia or alumina substrate. The package is equipped with large-diameter terminals to provide electrical access to the inside of the container. A lid or cover is soldered or welded to the top of the package to form a hermetically sealed assembly. It is designed to handle 40 to 80 watts of power per square inch, compared with 4 to 5 watts per square inch of power for a standard multichip hybrid package (MHP).

Prebake The preheating of an electronic assembly to drive out internal moisture or gases, or to stabilize the materials, prior to high-temperature operation such as soldering. Such prebaking also better ensures elimination of entrapped corrosive gas in sealed packages. *See also Burn-In, Stabilization Bake.*

Precious Metals Relatively scarce and valuable metals such as gold, silver, and platinum. Not always the same as noble metals. For instance, silver is a precious metal but not a noble metal. *See also Noble Metals.*

Precursor The polyacrylonitrile (PAN) or pitch fibers from which carbon graphite fiber is made. *Also called advanced composite. See also Organic Composite.*

Precursor Polymer A polymer that can be converted to the final desired polymer by a simple process such as heating or irradiation.

Prefiring The advance firing of thick-film circuits before firing of film resistors. The process higher-temperature firing of circuit materials without damage to resistors.

153

Preform (1) A pill, tablet, or formed material used in thermoset molding. The material is measured by volume, and the bulk factor of powder is reduced by pressure to achieve efficiency and accuracy. (2) A geometrically shaped, thin sheet of material such as solder or epoxy, used for soldering or bonding. The bond or glue line thickness is more precisely controlled using preforms. *See also Solder Preform.*

Preheating The heating of a compound prior to molding or casting in order to facilitate the operation, reduce cycle time, and improve the product.

Premix A molding compound prepared prior to and apart from the molding operations and containing all components required for molding, including resin, reinforcement fillers, catalysts, release agents, and other compounds.

Premolded Plastic Semiconductor Packages As opposed to molded plastic packages, electronic packages that do not involve direct molding of the chip. In these packages, the leadframe is molded into a plastic by transfer or injection molding. Chips are then secured in the molded package and wire-bonded to the appropriate leads. The package assembly is completed with no further molding operations. *See also Molded Plastic Semiconductor Packages, Transfer Molding.*

Prepolymers (1) A polymer in some stage between that of the monomers and the final polymer. The

molecular weight is, therefore, also intermediate. (2) As used in polyurethane production, a reaction product of a polyol with excess of an isocyanate.

Prepreg A ready-to-mold sheet that may be cloth, mat, or paper impregnated with resin and stored for use. The resin is partially cured to a B stage and supplied to the fabricator, who lays up the finished shape and completes the cure with heat and pressure. *See also B Stage.*

Preseal Visual Inspection A visual inspection of the entire hybrid circuit assembly—including components, wire bond interconnections, particulate, and any defects—prior to sealing the package.

Pressed Ceramic A ceramic part formed by applying pressure to the ceramic powder and binder prior to firing in a kiln. *See also Green Tape, Kiln.*

Pressed Powder *See Dry Pressing.*

Press-Fit Contact An electrical contact that is forced into a hole of an insulator- or conductor-type base material.

Press-Fit Pin A connector pin that is forced into a hole of a substrate or printed wiring board to form a seal without the use of solders or a weld.

Pressure Bag Molding A process for molding reinforced plastics in which a tailored flexible bag is placed over the contact lay-up on the mold,

154

sealed, and clamped in place. Fluid pressure, usually compressed air, is placed against the bag, and the part is cured. *Also called bag molding.*

Pressure Contact An electrical contact that is made by a spring-applied force.

Pretinned The application of solder to the leads of a component or to a wire prior to soldering the component or wire in place.

Primary Insulation The first layer of nonconductive material applied over a conductor, whose prime function is to act as electrical insulation.

Primary Side The side of a printed board that contains most of the components. *See also Secondary Side.*

Primer A material applied to a surface, prior to the application of a coating or adhesive, to improve the bond of the coating or adhesive to the surface.

Print and Fire In thick-film technology, the screening of the paste or ink on the substrate and the subsequent firing in a high-temperature furnace to a specific temperature profile.

Printed Board The general term for a completely processed printed circuit and printed wiring configuration, including single-sided, double-sided, and multilayer boards with rigid, flexible, and rigid-flex base materials.

Printed Circuit (1) A conductive pattern composed of printed components, printed wiring, or a combination thereof, that is formed in a predetermined arrangement on a common base. (2) Generically, a printed board that is produced by any of a number of techniques. *See also Printed Wiring.*

Printed Circuit Board A printed board that provides both point-to-point connections and printed components in a predetermined arrangement on a common base. *See also Printed Wiring Board.*

Printed Circuit Board Assembly An assembly that uses a printed circuit board for component mounting and interconnecting purposes. *See also Printed Wiring Board Assembly.*

Printed Circuit Laminates Laminates, either fabric- or paper-based in a resin or other matrix, that are covered with a thin layer of copper foil and used in the photofabrication process to make printed boards. *See also Laminate.*

Printed Contact The portion of a printed circuit, usually near the edge of the board, that provides the contact area for a connector. *Also called terminal area.*

Printed Wiring A conductive pattern that provides point-to-point connections, but not printed components, in a predetermined arrangement on a common base. *See also Printed Circuit.*

Printed Wiring Board (PWB) A printed board that provides point-to-point connections, but not printed components, in a predetermined arrangement on a common base. It can be single- or double-sided or a multilayer construction of either rigid or flexible composite materials. *See also Printed Circuit Board.*

Printed Wiring Board Assembly An assembly that uses a printed wiring board for component mounting and interconnection.

Printed Wiring Substrate The laminate base, or other base, onto which a printed wiring pattern is formed.

Printing The process of reproducing an image or pattern on the surface of a material by photoetching, vapor deposition, screen printing, diffusion, or other means.

Printing Parameters Factors that can be controlled in the screening process of thick films. Breakaway, speed, and pressure on the squeegee are examples. *See also Line Definition, Thick Film.*

Print Laydown In thick-film technology, the screening of a circuit pattern on a substrate.

Probe A rigid, pointed, wire-shaped metal device used for making electrical contact to a circuit pad for electrical test purposes.

Processor (1) In computer hardware, a unit that processes data. (2) In computer software, a program that compiles, assembles, and translates related functions for a specific language, including logic, memory, and control.

Procuring Agency A government, contractor, subcontractor, or other agency that contracts for the purchase of equipment, spare parts, services, or other items and has the authority to grant waivers, deviations, or exceptions to the procurement documents.

Production Lot Electronic assemblies manufactured on the same production line(s) by means of the same production techniques, materials, controls, and design. The production lot is usually date-coded to permit the control and traceability required for maintenance of reliability programs.

Production Master A one-to-one scale pattern that is used to produce rigid or flexible printed boards within the accuracy specified on the master drawing.

Profile (1) A graphic representation of time verses temperature of a continuous thick-film furnace cycle. (2) A graphic representation of the roughness, or flatness, of a substrate over some distance on the surface of that substrate.

Profilometer An instrument used to measure the degree of roughness of a surface. *See also Profile.*

Programmable Logic Array (PLA) An integrated circuit consisting of an array of combinational logic elements (circuits) with a fixed interconnection pattern in which connections can be made or broken after manufacture to perform specific logic functions. The PLA is typically a large set of AND gates driving several OR gates.

Programmable Via A via that is not on a prescribed grid pattern, as is a fixed via. *See also Fixed Via.*

Programmed Wiring A process in which wires are attached to a termination panel containing many posts by using programmable equipment.

Promoter *See Accelerator.*

Propagation The movement or travel of electromagnetic waves through a medium.

Propagation Delay The delay in time between the input and output of a signal. The delay time is measured in nanoseconds per foot of conductor length, and is dependent on both conductor length and the dielectric constant of the materials used with the circuitry. This is especially important in high-speed circuitry. *See also Dielectric Constant, High-Speed Circuitry.*

Propagation Speed A speed at which a wave travels through a medium. *See also Propagation Delay.*

Propagation Time The time required for a wave to travel between two points on a transmission line. *See also Velocity of Propagation.*

Propagation Velocity The speed at which a signal travels along a transmission line or through a medium.

Property Any physical, chemical, or electrical characteristic of a material.

Proportional Limit The greatest stress a material can sustain without deviating from the linear proportionality of stress or strain, according to Hooke's law.

Proton A positively charged particle equal to the negative charge of the electron but with a mass of 1846 times that of the electron.

Prototype A handmade working model representative of the final design that is used for evaluation and testing.

P-Type Semiconductor Material A crystal of pure semiconductor material to which an impurity, such as boron or gallium, has been added to give it a deficiency of electrons. The majority carriers are holes. *See also Hole, N-Type Semiconductor Material.*

Pull Strength *See Bond Strength.*

Pull Test A test often used to measure the bond strength of a lead, wire, or conductor.

Pulse A sudden change in current or voltage from one value to a higher or lower value and back to its original value in a specific time. *See also Impulse.*

Pulse Duration The time interval between a reference point on the leading edge of a pulse waveform and a reference point on the trailing edge of the same waveform. The two reference points are usually 90 percent of the steady-state amplitude of the waveform existing after the leading edge, measured with respect to the steady-state amplitude existing before the leading edge. If the reference points are 50 percent, the symbol t_w and the term *average pulse duration* should be used. *Also called pulse time.*

Pulse Vacuum *See Solder Extraction.*

Pultrusion Reversed "extrusion" of resin-impregnated roving in the manufacture of rods, tubes, and structural shapes of a uniform cross section. The roving, after passing through the resin dip tank, is drawn or pulled (hence, pultrusion) through a die to form the desired cross section and subsequently through a heater to cure the composite. *See also Extrusion.*

Puncture An undesirable hole or dislodged volumetric section in an insulating material as a result of a high-voltage discharge. *See also Impulse.*

Puncture Strength The voltage at which the dielectric or base materi-al is punctured by the voltage stress at the point of puncture. Usually, the voltage level at puncture is not known, and steps must be taken to eliminate the basic cause of the stress. *See also Partial Discharge.*

Purge To evacuate a chamber, containing components or electronic assemblies, of moisture and other contaminates prior to sealing or backfilling with an inert gas.

Purple Plague A purple-colored compound that is formed in bonding gold to aluminum. The gold-aluminum compound is activated by moisture, by temperatures exceeding 350°C and by the presence of silicon. It causes serious degradation of semiconductor devices.

Push-Off Strength The force required to break the bond between a chip and its mounting pad, as measured by applying a force to one side of the chip and parallel to the mounting pad. The bonding area should be free of excessive fillets in order to get an accurate reading.

Pyrolyzed Pertaining to a material that has undergone chemical decomposition by heat, usually without oxidation.

Pyrophosphate Copper Copper that is electrodeposited using a copper pyrophosphate electrolyte bath.

Q

Q Factor The relationship between stored energy and rate of dissipation. (1) For an inductor, the ratio of coil reactance to effective coil resistance at any given frequency. (2) For a capacitor, the ratio of susceptance to effective shunt at any given frequency. (3) For a magnetic or any dielectric material, 2 pi times the ratio of maximum stored energy to energy dissipated in the material per cycle.

Quad Flat Pack (QFP) A chip carrier made of ceramic or plastic that has its leads bent downward and away from the four sides of the package. Generally refers to a plastic quad flat package built to Electronics Industry Association of Japan (EIAJ) standards. *See also Plastic Quad Flat Package.* See Fig. 15.

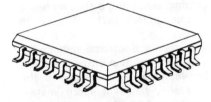

Figure 15: Quadpack

Quad In-Line Package (QUIP) A plastic package similar to a dual in-line package with leads extending out of the package on 1.27 mm centers. Half of the leads are bent at the edge of the package, while the remaining project past the edges of the package 1.27 mm before being bent down, creating a staggered lead configuration. The package can also have staggered leads to halve the effective center-to-center spacing and can be surface or through-hole mounted.

Quadpack *See Quad Flat Pack.*

Qualified Products List (QPL) A list of commercial products and materials that have been pretested and found to meet prescribed requirements, especially government specifications.

Quality The conformance of a component system to a specification or to a customer requirement.

Quality Assurance A prescribed set of quality control and quality engineering methods used to ensure the manufacture of a product of acceptable quality standards.

Quiescent Voltage or Current The DC voltage or current at a terminal with reference to a common terminal, normally grounded, when no signal is applied.

159

R

Rack and Panel Connector A connector that, when fully inserted, joins the inside back end of a cabinet with the drawer containing the electrical equipment.

Radar Equipment or systems for radio detection and ranging.

Radar Signature Radar-detected patterns or features that are unique to a given target and hence can be used to identify that target.

Radial Spread Coating See Glob-Top Coating.

Radiation The emission, transfer, or propagation of energy in the form of electromagnetic waves or particles through space. Radiation can be thermal, nuclear, or electromagnetic. See also Electromagnetic Radiation, Gamma Radiation.

Radiation Hardened (1) A process by which components and circuits are preexposed to high gamma and neutron radiation so that their performance is not degraded when they are subsequently exposed to hostile radiation environments. (2) The design and manufacture of components and circuits such that their performance is not degraded beyond usable limits when they are exposed to gamma and neutron radiation. Also called radiation resistant.

Radiation Resistant See Radiation Hardened.

Radiation Transfer Index (RTI) In fiber optics, the transmission performance of an optical fiber cable, including the coupling and propagation losses.

Radio Frequency (RF) Electrical frequencies in the range of 30 kHz to 300,000 mHz.

Radio Frequency Interference (RFI) Electrical signals from internal or external sources that interfere with the operation of an electrical system or electronic equipment.

Radiograph An X-ray photographic image of the inner parts of an object, such as a sealed electronic package, a metallized hole in a ceramic substrate, a solid material, or a component part.

Radio Waves Electromagnetic radiation in that portion of the spectrum between audio and infrared, in the frequency range of approximately 10 kHz to 100,000 mHz. See also Electromagnetic Radiation.

Random-Access Memory (RAM) A memory in which information can be independently stored or retrieved. Usually bits of information are only stored temporarily.

Random Failure A failure that occurs on a random or unpredictable basis, with the failure rate for the sample population remaining nearly constant. *Also called random network.*

Random Network *See Random Failure.*

Rapid Impingement Speed Plating (RISP) A proprietary high-speed plating process accomplished by using high current density and forced movement of the plating solution. Both pushing and pulling of the solution are utilized. *See also Electroplating, Plating.*

Rated Temperature The maximum temperature at which an electric component can operate for extended periods without undue degradation or safety hazard. As defined by the Institute of Electrical and Electronics Engineers under IEEE-1-1986, temperature class refers to the ability to give a specified performance for a specified length of time, usually 20,000 hours.

Rated Voltage The maximum voltage at which an electrical system or component can operate safely for extended periods of time without excessive degradation.

Rating The nominal value of any electrical, thermal, mechanical, or environmental quantity assigned to define the operating conditions under which a component, machine, apparatus, or device is expected to give satisfactory service. *See also Maximum Rating.*

RC Network A network composed entirely of resistors and capacitors.

Reactance (X) The opposition offered to the flow of alternating current by the inductance or capacitance of a component or circuit.

Reaction Etching A chemical process in which metallic circuits are formed by removing uncoated areas of a conductive pattern. The chemical dissolves the undesired metal, usually copper.

Reaction Injection Molding (RIM) A plastic molding process in which proportional amounts of two reactive materials are fed into a reaction chamber, thoroughly mixed together, and subsequently forced into a mold for polymerization or cure.

Reactive Diluent A diluent compound, containing one or more epoxy groups, that functions mainly to reduce the viscosity of the mixture but is also part of the polymerization reaction. *See also Diluent.*

Reactive Ion Etching (RIE) A plasma-etching process using a relatively low pressure and high electric field, in which material is removed primarily by chemical reaction and active radicals. Some material may also be removed physically by ion bombardment. A mask is usually used in order to remove only selected areas.

Reactive Metal Any metal that readily reacts to form chemical compounds. Examples are copper and iron.

Read-Only Memory (ROM) A memory device in which information is permanently stored during its manufacture or installation and cannot be erased.

Rebonding over Bond (1) To apply a second bond on the same pad area after the first, damaged bond has been removed. (2) To place a second bond adjacent to the initial bond.

Receiver A system that converts electrical waves into audio or video form.

Receiving Element In fiber optics, the accepting side of the termination at the conductor interface.

Receptacle Connector A connector that is designed for easy mounting on a panel, bulkhead, or chassis and mates with a plug connector. *See also Chassis, Panel.*

Reducing Atmosphere An atmosphere to which a reducing gas such as hydrogen or nitrogen, or some gas combination, has been added to prevent oxidation of metal parts while they are being fired or otherwise processed. *See also Oxidation.*

Reduction Dimension An exact dimension that is placed on an actual-size layout drawing, located between two marks, and used to verify the exact distance after photographic reduction has been made.

Redundancy A design that employs additional, duplicate components or circuits, more than are necessary to perform the function. The object is to improve reliability, since the duplicate part will assume the function of the original part once the original part has failed.

Reference Edge An edge from which measurements are made or a conductor can be identified with the number 1 conductor being the closest to the edge.

Refiring In thick-film technology, the recycling of a film resistor through the firing cycle in order to change the resistor value to the finally desired value.

Reflow The application of heat to a surface containing a thin deposition of a low-melting metal or alloy, resulting in the melting of the deposition, followed by solidification. A common example is the reflow of deposited solder metal or solder paste. *See also Reflow Soldering, Solder Paste.*

Reflow Soldering A method in which solder metal or paste is first applied to the surfaces of the parts to be joined and subsequently heated, causing the solder to melt and reflow to form a solder joint. *See also Paste Soldering, Solder Paste.*

Refractive Index (1) The ratio of the velocity of light in a vacuum to its velocity in a substance. (2) The ratio of the sine of the angle of incidence to the sine of the angle of refraction.

Refractory Metal A metal such as molybdenum or tungsten that has an extremely high melting point. As an example, molybdenum has a melting point of 2620°C.

Registration The alignment of a circuit pattern on a substrate or base pattern with respect to other layers of a double-sided or multilayer board. This ensures proper alignment of conductors from layer to layer, and the reliable interconnection between layers. *See also Multilayer Board.*

Registration Marks The marks used for aligning artwork or successive layers of printed wiring patterns on multilayer boards. *Also called alignment marks, corner marks, or fiducial marks.*

Regrind In a thermoplastic molding process, the excess or waste material that can be reground and mixed with virgin raw material, within limits, for molding future parts.

Regulation The absolute change in a parameter for a change of a circuit variable from one level to another. The change is usually normalized as a percentage but may not always be normalized.

Reinforced Molding Compound A plastic to which glass, cotton, or other fibrous material has been added to improve certain physical properties, such as flexural strength.

Reinforced Plastic A plastic with strength properties greatly superior to those of the base resin, resulting from the presence of reinforcing materials, such as glass fibers, in the composition.

Reinforced Thermoplastics Reinforced plastic molding compounds in which the plastic is thermoplastic.

Relative Humidity The ratio of the quantity of water vapor present in the air to the quantity that would saturate it at any given temperature.

Relay A device that is electrically controlled to mechanically open and close electrical contacts in the same or another circuit.

Release Agent *See Mold Release.*

Release Paper An impermeable paper film or sheet coated with a material that prevents adhering of the sheet to the prepreg. Commonly used as separate sheets in the storage of prepregs. *See also Prepreg.*

Reliability (1) The probability that a system or component will have a failure-free performance under certain environmental conditions for a period of time. (2) The continued conformance of a device or system to a specification over an extended period of time.

Reliability Assurance A technology that assesses the reliability of a product by means of surveillance and measurement of the factors of design and production that affect it.

Reluctance The measure of a material's resistance to the passage of magnetic flux.

Rents Rule A mathematical relationship which states that the number of input or output terminations in a logic package is proportional to a fractional power of the number of logic gates interconnected in the package. The functional power can vary with circuit complexity.

Repair A manufacturing operation that restores a part or an assembly to an operable condition but does not eliminate nonconformance. *See also Rework.*

Resin A high-molecular-weight organic material with no sharp melting point. For general purposes, the terms *resin, polymer,* and *plastic* are often used interchangeably. *See also Plastic, Polymer.*

Resin Bath In the manufacture of fiber glass reinforced plastics, a resin-filled container in which the reinforcing glass fiber materials are immersed and wetted with resin. This is a common process for the manufacture of epoxy-glass and other laminates. The resin-soaked glass fabric is partially cured to a prepreg state. *See also Laminate, Prepreg.*

Resin Desmearing The removal of unwanted smeared resin by an etching operation, such as chemical etching or plasma etching. *See also Plasma Etching.*

Resin-Rich Area The location in a printed board of a significant thickness of nonreinforced surface-layer resin that is the same composition as the resin within the base material. *See also Resin-Starved Area.*

Resin Smear The unwanted movement of a resin material to a conductive surface, such as the transfer of epoxy resin from a nonconductive layer to the exposed conductive layer in drilled holes of printed wiring boards. The resin smear results in an undesirable covering of conductor layers in the hole, so that through hole plating cannot be properly achieved. The resin is removed by desmearing operations in order to achieve properly plated interconnections in the hole. *See also Resin Desmearing.*

Resin-Starved Area The location in a printed board that does not have a significant amount of resin to completely wet out the reinforcing material. It is evidenced by low-gloss dry spots or exposed fibers. *See also Resin-Rich Area.*

Resist A material such as ink or paint that is used to protect selected areas during chemical etching, plating, and soldering. Examples are solder resist, plating resist, and photoresist.

Resistance The property of an electrical conductor that determines the amount of current produced by a given difference of potential. The ohm is the unit of resistance.

Resistance Soldering A process in which an electrical current is passed through an electrode and creates heat. The heated electrode is then placed in contact with the solder, which melts. *See also Reflow Soldering, Soldering, Wave Soldering.*

Resistance Welding The joining of two electrically conductive materials by heat and pressure. The heat is generated by passing electrical current through the two conductors, which are held together mechanically until the weld is complete. *See also Welding.*

Resistivity The characteristic of a material that resists passage of electric current either through its bulk or on its surface. The unit of volume resistivity is the ohm-centimeter, and the unit of surface resistivity is ohm/square. *See also Insulation Resistance, Sheet Resistivity, Surface Resistivity, Volume Resistivity.*

Resistor An electrical component made of material that has a known resistance and opposes the flow of electrical current. *See also Chip Resistor, Component, Thick-Film Resistor.*

Resistor Drift The change in resistor value of a resistor that takes place over a period of time through aging. It is rated as a percentage change per 1000 hours.

Resistor Geometry In thick films, the physical outline of a screened resistor. *See also Thick-Film Resistor.*

Resistor Overlap In screen printing, the overlap contact area that the film resistor makes with a film conductor. *See also Screen Printing, Thick Film.*

Resistor Termination *See Resistor Overlap.*

Resistor-Transistor Logic (RTL) A logic circuit composed of several resistors, a transistor, and a diode.

Resolution *See Line Definition.*

Reverse Bias An external voltage applied to semiconductor P-N junction to reduce the current across the junction. Opposite of forward bias. *See also Forward Bias.*

Reverse Current, DC The DC current through a semiconductor diode in the reverse direction. *See also Forward Current.*

Reverse Image The pattern of resist on a printed board that is needed to allow the exposure of conductive areas for subsequent plating.

Reverse Voltage, DC The DC voltage applied to a semiconductor diode that causes current in the reverse direction. *See also Forward Voltage.*

Reversion A chemical reaction in which a polymerized material partially or completely degenerates to a lower polymeric state or to the original monomer. It is caused by

165

heat or moisture, or a combination of both, and results in significant changes in physical and mechanical properties. *See also Hydrolytic Stability, Hydrophilic, Hydrophobic, Polymer Reversion.*

Rework (1) A manufacturing operation that restores a part or an assembly not only to an operable condition but to the requirements of the contract, drawings, or other specifications. (2) The replacement of a device, which includes the removal of the device, preparation of the pad or area for the new component, mounting, and interconnections. Rework is sometimes required for either replacing a defective component or incorporating an engineering change. *See also Repair.*

RF Connector A connector to which a coaxial cable can readily be attached. It exhibits the same characteristics as the cable. *See also Coaxial Cable, Radio Frequency.*

Rheology The study of the change in form and flow of matter, such as plastic resins, embracing elasticity, viscosity, and plasticity. *See also Viscosity.*

Rib A structurally reinforcing member of a formed part or a molded plastic part.

Ribbon Cable A flat cable whose conductors are insulated from each other in individual jackets. Hence, ribbon cable does not have a smooth surface, as does flat cable. *See also Flat Cable.*

Ribbon Interconnect An electrical interconnection between circuits or to the output pins of an electronic package in which flat, narrow ribbon conductors (e.g., flat cable, ribbon cable, ribbon wire) are used. *See also Flat Cable, Ribbon Cable, Ribbon Wire.*

Ribbon Wire A flat, flexible metal wire having a rectangular cross section. *See also Round Wire.*

Rigid Coating A conformal coating, usually a thermosetting resin, that does not contain any plasticizers to keep the coating pliable. *See also Conformal Coating.*

Rigid-Flex Printed Board A printed board that consists of both rigid and flexible base materials. Normally, the flexible portion is bonded directly into the rigid portion. *See also Flexible Printed Wiring, Printed Circuit Board.*

Rigidsol Plastisol having a high elastic modulus, usually produced with a crosslinking plasticizer. *See also Plastisol.*

Riser Pin An insulated metal pin that extends only on the inside of a package, providing a method of interconnection between multiple substrates.

Riser *See Via.*

Rise Time The time required for the initial edge of a pulse to rise from 10 percent to 90 percent of its peak value.

Robotic Pertaining to mechanical devices that are designed to perform a variety of tasks formerly done by human beings.

Rockwell Hardness Number A number derived from the net increase in depth of impression as the load on a penetrator is increased from a fixed minimum to a higher load and then returned to minimum. Penetrators include steel balls of several specified diameters and a diamond-cone penetrator.

Roentgen The amount of radiation that will produce one electrostatic unit of ions per cubic centimeter of volume.

Roller Mill A jarlike container, usually ceramic, that is partially filled with hard pebbles. Mixtures of powders and liquids are poured into the mill and the mill is rotated for predetermined time periods. This process mixes the powder and liquids into a very uniform slurry. *Also called pebble mill. See also Green Tape, Slurry, Tape Casting.*

Root Mean Square (rms) The square root of the average of the sum of the squares of a series of values.

Rosin A hard, natural resin consisting of abietic and primaric acids and their isomers, some fatty acids, and terpene hydrocarbons. Rosin is extracted from pine trees and subsequently refined.

Rosin-Activated Flux *See Flux, Rosin-Activated; and other listings under Flux.*

Rosin Flux *See Flux, Rosin; and other listings under Flux.*

Rosin Joint *See Rosin Solder Connection.*

Rosin Mildly Activated *See Flux, Rosin Mildly Activated; and other listings under Flux.*

Rosin Solder Connection A defective solder joint in which the connection is held together by an invisible film of flux. *Also called rosin joint.*

Rotational Casting A method used to make hollow articles from thermoplastic materials. Material is charged into a hollow mold capable of being rotated in one or two planes. The hot mold fuses the material into a gel after the rotation has caused it to cover all surfaces. The mold is then chilled and the product stripped out. *Also called rotational molding.*

Rotational Molding *See Rotational Casting.*

Round Wire Any circular conductor, such as magnet wire or hookup wire. *See also Hookup Wire, Magnet Wire.*

Routing Program An automated interconnecting layout program.

Roving A collection of bundles of continuous-fiber filaments, either as untwisted strands or as twisted yarns. Rovings may be lightly twisted, but for filament winding they are generally wound as bands or tapes with as little twist as possible. *See also Strand, Yarn.*

Rubber An elastomer such as natural rubber, capable of rapid elastic recovery. *See also Elastomer.*

Rubylith A laminate material consisting of a thin red film with a heavier clear backing. It is used to make master artwork by cutting and peeling away portions of the red layer.

Runner The secondary material feed channel in an injection or transfer mold that runs from the inner end of the sprue to the cavity gate. *See also Gate, Sprue.*

Runners In plastic molding, all the channels through which molten or liquid plastic raw materials flow into the mold. *See also Sprue.*

S

Sandwich Construction A panel consisting of some lightweight core material that is bonded to strong, stiff skins on both faces. *See also Honeycomb.*

Saponifier An alkaline chemical that is added to water to clean up rosin-base flux residue.

Sapphire The monocrystalline form of alumina, Al_2O_3. An insulating material on which silicon can be grown and etched away to form a solid-state device, sometimes described as silicon on sapphire.

Scaling The separation of film conductors and resistors from the substrate.

Scallop Marks Screen-printed lines having irregular or jagged edges, which can be caused by incorrect squeegee pressure, insufficient emulsion thickness, or incorrect screen mesh size. *See also Screen Printing.*

Scarf Joint A joint made by cutting away similar angular segments of two adherends and bonding the adherends with the cut areas fitted together. *See also Adherend.*

Scavenging *See Leaching.*

Schematic Diagram A drawing showing only the graphic symbols, electrical connections, and components that make up a circuit.

Schottky-Barrier Charge-Coupled Device A buried-channel charge-coupled device that uses a Schottky barrier junction to isolate the transfer gate.

Schottky Diode A diode formed by depositing a metal film on a semiconductor surface of sufficiently high resistivity to form an energy barrier. A Schottky diode is sometimes used as part of a bipolar transistor structure.

Scored Substrate A substrate that has been diamond-scribed to form thin cut lines for subsequent separation of the substrate into smaller parts by simple breaking. *See also Scribe.*

Screen A metal or fabric network, of various size squares or mesh, that is mounted snugly on a frame. An emulsion is then bonded to the screen, and circuit patterns and configurations are superimposed by photographic means. *See also Photolithography, Screen Mesh, Screen Printing.*

Screenability The qualitative characteristic of a solder or thick-film paste with respect to its ease of screen printing, which depends on material properties. *See also Screen Printing, Thick-Film Technology.*

Screen Deposition *See Screen Printing.*

Screen Frame A rectangular configuration made of wood, metal, or plastic on which a screen is snugly mounted and held firmly in place. *See also Screen.*

Screen Mesh The woven fibers or metal wires that support the emulsion and allow the paste or ink to flow through during screen printing. The diameter and spacing of the fibers determine the size of the mesh. *See also Screen, Screen Printing.*

Screen Printing A thick-film process in which a paste or ink is squeezed through open areas of a screen and transferred to the surface of a substrate to form film circuits and configurations. *Also called screen deposition or silk screening. See also Screen, Thick-Film Technology.*

Scribe To scratch with a hard, pointed material such as a diamond. *See also Scored Substrate.*

Scribe and Break A technique in which a hard, pointed material such as a diamond is used to scratch a ceramic substrate that is subsequently put under tension to break along the scratched line. *See also Scored Substrate.*

Scrubbing To apply circular action to a clean chip or substrate during the chip- or die-bonding operation to improve the wettability of the bonding area, and hence form a stronger bond. *Also called scrubbing action. See also Die Bond.*

Scrubbing Action *See Scrubbing.*

Sealed Chips on Tape (SCOT) Integrated circuit chips mounted on tape-supported leads and sealed, usually with a blob of adhesive, while still in the tape and reel format.

Sealing In hybrid operations, a process in which the lid or cover is joined to the header to form a sealed electric package. *See also Hybrid Electronic Assembly.*

Sealing Plug (1) A plug that is placed in a contact opening of an unused connector insert to provide an environmental seal. (2) A metal disk that is bonded to the lid of an electronic package after the package has been gas filled or evacuated, as is sometimes done in hybrid electronic assemblies. *See also Hybrid Electronic Assembly.*

Search Height The height of the bonding tool above the bonding area at which final adjustments in the location of the bonding area under the tool are made prior to lowering the tool for bonding. *See also Bonding.*

Secondary Side The side of a printed wiring board or a packaging and interconnecting structure opposite the primary side. It is the side on which the soldering is performed when through-hole mounting technology is employed. *See also Primary Side.*

Second Bond The second bond of a bond pair made to form a conductive connection. *See also Wire Bond.*

Second Radius The radius of the back edge of the bonding tool foot. *See also Bond Tool.*

Selective Etching A process in which the etching is restricted by using a chemical that will attack only one of the exposed metals.

Selective Plating The electrochemical deposition of a metal on specific areas of a part. The other areas are covered with a masking or resist-type material prior to plating. *See also Electroplating.*

Self-Extinguishing The ability of certain materials, especially plastics, to extinguish their own flame after the source of the flame has been removed. This characteristic is usually imparted to plastics by the presence of certain elements, such as chlorine or fluorine, in the molecule, or by the addition of flame retardants to the plastic.

Self-Healing A characteristic of certain gel-type polymers, such as silicone gel, to reflow into a homogeneous mass once a test probe has been removed. *See also Silicones.*

Self-Heating The generation of heat caused by a chemical or exothermic reaction.

Self-Passivating Glaze A glossy finish that appears on the surface of thick-film resistors after firing and seals the surface from moisture absorption. *See also Thick-Film Resistor.*

Self-Stretching Soldering Technology (SST) A technique used to increase the height of a solder joint. When two different solder alloys have different thermal expansion characteristics or different-size solder bumps, the greater expansion of the nonfunctional bumps increases the height of the functional solder joint. This technique produces higher-functioning solder joints that can withstand greater thermal stress without cracking.

Semiadditive Process   An additive process in which the entire thickness of electrically isolated conductors is obtained by the use of electroless metal deposition and electroplating, etching, or both. *See also Fully Additive Process, Subtractive Process.*

Semiconductor   A material whose electrical conductivity and resistivity lie between those of a conductor and an insulator. Typical materials are germanium, lead sulfide, silicon, gallium arsenide, and silicon carbide. *See also Semiconductor Device.*

Semiconductor Carrier   A protective structure that is used for mounting semiconductor devices and has metallized internal pads and external feedthroughs for connecting the chip to a substrate or printed wiring board. *See also Semiconductor Package.*

Semiconductor Chip   A square or rectangular piece of semiconductor material that has been processed to form an electrical device such as a transistor or an integrated circuit. *See also Semiconductor Device.*

Semiconductor Device   A device whose essential characteristics are governed by the flow of charge carriers within a semiconductor.

Semiconductor Diode   A semiconductor device having two electrodes and exhibiting a nonlinear voltage-current characteristic.

Semiconductor, Integrated Circuit   (1) A synonym for monolithic semiconductor integrated circuit. (2) An integrated circuit consisting of elements formed in situ on or within one or more semiconductor substrates with at least one of the elements formed within the substrate(s).

Semiconductor Package   An enclosure for one or more semiconductor chips that allows electrical connection and provides mechanical and environmental protection. *Also called IC package.*

Sensitivity   The ratio of the change in a parameter to a change in a circuit variable other than temperature. The change in the parameter may or may not be normalized to a reference value of the parameter. The ratio is usually the average value for the total change of the circuit variable.

Separating Force   *See Engaging and Separating Force.*

Separator Ply   *See Shear Ply.*

Sequentially Laminated Multilayer Printed Wiring Board   A multilayer board composed of double-sided boards and/or multilayer boards that are bonded together, with each containing plated-through-hole interconnections prior to bonding and having blind and/or buried vias.

Serpentine Cut A zigzag cut in a film resistor, made in the trimming process, to increase the resistor length, thereby increasing the resistance of the resistor. *See also Film Resistor.*

Serrations Miniature grooves on the inside diameter surfaces of a terminal wire barrel. These provide additional electrical conductivity and improved tensile strength after crimping.

Service Rating The maximum current, voltage, or temperature that an electrical component is designed for and is capable of withstanding on a continuous basis.

Servomechanism An automatic device that uses feedback to control systems, usually by inserting, at the input, control developed from samples of the output.

Set (1) In a mechanical deformation, the strain remaining after complete release of the load producing the deformation in a material. *See also Strain, Stress.* (2) In polymers, to convert an adhesive or other resin form into a fixed or hardened state by chemical or physical action, such as condensation, polymerization, oxidation, vulcanization, gelation, hydration, or evaporation of volatile constituents. *See also Polymer.*

S Glass Glass fabric made with fibers of very high tensile strength, to meet high-performance strength requirements.

Shadow Effect During wave soldering, as the assembly moves over the surface of the molten solder, the creation of a void, depression, or skip behind a large component so that solder does not reach the solder pad.

Shape Factor For an elastomeric slab loaded in compression, the ratio of the loaded area to the force-free area. *See also Elastomer.*

Shear An action or stress resulting from applied forces that causes two contiguous parts of a body or two bodies to slide relative to each other in a direction parallel to their plane of contact.

Shear Ply A low-modulus layer, made of rubber or adhesive, that is interposed between a metal and composite to control differential shear stresses.

Shear Rate The relative rate of flow of viscous fluids. *See also Thixotropic, Viscosity.*

Shear Strength The maximum shear stress that a material is capable of sustaining. In testing, the shear stress is caused by a shear or torsion load and is based on the original specimen dimensions.

Sheet A flat piece of plastic or metal, usually having a thickness of over 10 mils (0.01 inch) and commonly made by extrusion, calendaring, casting, or rolling. *See also Film, Foil.*

Sheet Molding Compound (1) Compression molding material consisting of glass fibers longer than ½ inch and thickened polyester resin. Possessing excellent flow, it results in parts with good surfaces. (2) A composite of glass fibers, resins, pigments, fillers, and other additives that have been compounded and processed into sheet form to facilitate handling in the molding operation.

Sheet Resistivity The electrical resistance across the surface of a square of screened and fired thick-film paste or other resistive film material. This resistance characteristic is the measured resistance between opposite edges of a unit square of any dimension. Since the same value exists in both directions of the square, and since the length divided by the width is a dimensionless number known as aspect ratio and arbitrarily called a square, the unit of sheet resistivity is ohms per square. It is similar to surface resistivity. Thick-film pastes are characterized by their sheet resistivity value. *See also Aspect Ratio, Surface Resistivity, Volume Resistivity.*

Shelf Life (1) The length of time that a molding compound or other resin form can be stored without losing any of its original physical or molding properties. (2) The useful storage life of the raw materials or

mixture in a polymeric formulation. *See also Polymer.*

Shell The outermost case or enclosure that houses the insert and contacts of a connector and provides correct alignment and protection of projecting contacts.

Shielding Effectiveness (SE) The effectiveness of a shielding barrier, such as a metallic or semimetallic covering, in a measurement of the reduction in field strength between a source and a receptor of electromagnetic or radio frequency energy. *See also Electronic Shielding.*

Shore Hardness Test A test that measures the surface hardness of soft and semihard materials, such as rubber and plastics using a durometer. Shore designations are given to tests made with a specified durometer instrument, such as Shore A and Shore B. *See also Durometer.*

Short *See Short Circuit.*

Short Circuit A disrupted circuit of relatively low resistance whose current is flowing in an undesired path. *See also Open Circuit.*

Shorting Plug *See Dummy Connector Plug.*

Shrinkable Tubing *See Heat Shrinkable, Solder Sleeve.*

Shrinkage The decrease in the physical dimensions of a plastic molded part through cooling, or of a polymer upon polymerizing. The former

173

is called *mold shrinkage* and the latter is called *polymerization shrinkage*. *See also Polymer.*

Shrink Quad Flat Package (SQFP) A quad flat package with a lead pitch of 0.016 inch or less.

Shrink Small Outline Integrated Circuit (SSOIC) A plastic package with gull wing leads on two sides and a lead pitch of 0.025 inch or less.

Shroud *See Insulation Support.*

Shunt An electrical device, made of low-resistance material, that is used to deliberately bypass part of a circuit.

Signal An electrical impulse of a predetermined voltage, current, polarity, and pulse width.

Signal Distribution The conductors within an electronic package that interconnect the drivers and receivers.

Signal Processing The treatment or manipulation of a signal by some means. *See also Signal.*

Signal-to-Noise Ratio The ratio of the root mean square (rms) signal power to the rms noise power at the output of a radar receiver.

Signal Wiring The conductors that carry the electronic signals.

Silanes Silicon-based chemical compounds used to treat inorganic materials such as glass fibers or mineral fillers, thereby improving the adhesion between these materials and organic resins. Silanes may be applied to the inorganic materials, added to the organic material, or both. One important use is for adhesion-promoting coupling agents between glass fibers and organic resins in the manufacture of laminates for printed wiring boards. *See also Coupling Agent.*

Silicon A semiconductor material; the base material for most semiconductors devices. *See also Semiconductor, Semiconductor Device.*

Silicones Resinous materials derived from organosiloxane polymers furnished in different molecular weights. This group includes liquid gel and solid resin forms used in applications such as thermosetting and elastomeric encapsulation resins, greases, and heat-stable fluids and compounds.

Silicon Monoxide An insulating material that is vapor-deposited on selected areas of thin-film circuitry, such as on a semiconductor chip. *See also Passivation.*

Silicon on Sapphire A semiconductor device formed on a sapphire substrate. *See also Sapphire, Semiconductor, Semiconductor Device.*

Silicon-on-Sapphire (SOS) Technology The technology whereby monocrystalline films of silicon are epitaxially deposited onto a single-crystal sapphire substrate to provide the basic structure for the fabrication of dielectrically isolated active and/or passive elements.

Silk Screen (1) A screen of closely woven silk fiber mesh that is stretched over a metal or wooden frame and is used to hold an emulsion outlining a circuit pattern. It is used in thick-film screen printing of film circuits and components. (2) Any type of screen, such as stainless steel or nylon, used in a printing operation.

Silk screening *See Screen Printing.*

Silver A metal with good electrical conductivity and corrosion resistance. It is relatively inexpensive and is used to electroplate copper conductors, making them readily solderable. Silver can dissolve into the solder, however, causing a solder joint of poor reliability. This mechanism is called *solder leaching*. *See also Solder Leaching.*

Silver Migration The ionic removal of silver and its redeposition in an adjacent area under the influence of migration-inducing conditions.

Simulated Aging Accelerated aging of a material under conditions of both high and low temperatures and humidity in an attempt to produce changes in its properties that are similar to those changes that occur during extended exposure to normal environmental conditions. *See also Accelerated Aging, Burn-In, High Temperature Reverse Bias Test.*

Single Chip Carrier An enclosure or electronic package that houses one chip and connects the chip terminals to the next higher level. *See also Semiconductor Package.*

Single Chip Module (SCM) A completed module that contains only one chip. *See also Module.*

Single In-Line Package (SIP) A package resembling a dual in-line package except that it contains only one line of leads instead of a double line of leads. *See also Dual In-Line Package.*

Single-Layer Alumina Metallized (SLAM) Package A leadless package that does not have a cavity. It is made of alumina ceramic and sealed to a ceramic cap with a ceramic or glass bond.

Single-Sided Substrate A packaging and interconnecting assembly, such as a printed wiring board, with components mounted on the primary side only. *See also Double-Sided Substrate, Primary Side.*

Sinking The undesirable electrical shorting of one conductor to another in the screen printing of multilayer circuits. It is caused by downward movement of the top conductor through the molten glass at the crossovers. *See also Screen Printing.*

Sink Mark A depression or dimple on the surface of an injection-molded plastic part. It is caused by the collapse of the surface following local internal shrinkage after the gate seals. *See also Gate.*

Sintering A process of bonding metal or other powders together using pressure and subsequently firing them into a strong cohesive mass.

Sizing Agent A chemical treatment containing starches, waxes, or other materials that is applied to fibers, making them more resistant to breakage during the weaving process. The sizing agent must be removed after weaving, as delamination and moisture pickup problems would result if it remained in the final laminate made from woven fiber.

Skew Ray In fiber optics, a ray that does not intersect the axis of a fiber but rather travels around the fiber along its length and outside surface.

Skin Effect The increase in resistance of a conductor at microwave frequencies that is caused by the tendency of electric current to concentrate at the conductor surface.

Slice *See Master Slice, Wafer.*

Slip Casting A ceramic fabrication process in which a low-viscosity aqueous-based ceramic slurry is poured in a plastic mold. The mold draws water from the slurry in contact with it, thus leaving a deposited layer of solid ceramic material on the wall of the mold. When a sufficiently thick layer of material

has been built up, the liquid remaining in the mold is drained and the ceramic part is dried and further processed and baked in a predetermined processing cycle. *See also Dry Pressing, Slurry, Tape Casting.*

Slipping The lateral movement of tensioned fiber on a composite surface to a new, unanticipated fiber angle. *See also Composite.*

Slip Rings Electromechanical devices used to transfer current between members that possess relative rotary motion. *See also Brushes.*

Slump The spreading of a screen-printed thick-film paste after printing and prior to drying, resulting in a loss of line definition. It is generally caused by low viscosity of the paste. *See also Screen Printing.*

Slurry A relatively thick mixture of solids, in suspension, in a liquid. *See also Green Tape, Roller Mill.*

Slush Molding A method for casting thermoplastics in which the resin, in liquid form, is poured into a hot mold, where a viscous skin forms. The excess slush is drained off, the mold is cooled, and the molded hollow part is stripped out.

Small Outline Integrated Circuit Package (SOIC) A plastic integrated-circuit package for surface-mount applications that has leads on two opposite sides.

Small Outline J-Leaded Package (SOJ) A small outline integrated-circuit package with J leads arranged in line and defined by the width of the plastic body.

Small Outline Package (SOP)   A small, rectangular, integrated-circuit surface-mount electronic package with leads on two sides and with 1.27 mm, 1.0 mm, or 0.85 mm spacing.

Small-Scale Integration (SSI)   A measure of complexity describing single microcircuits that contain fewer than 12 equivalent gates or circuitry of comparable complexity. *See also Large-Scale Integration, Medium-Scale Integration.*

Smart Card   A thin plastic card similar to a wallet-size credit card with a computer chip embedded in it. The card can be inserted into readers or scanners for direct transactions with a bank account or other account.

Smart Connector   A connector having an active electrical device in its construction that allows the connector to react to signals to which it would normally not react.

Smart Materials   Materials that will sense changes in desired parameters and react to or control such change in the desired manner. One such example is smart skins. *See also Smart Skins.*

Smart Skins   Smart materials used to control thermal conductivity by the outer skin of equipment, such as on military aircraft and tanks. *See also Smart Materials.*

Smeared Bond   A wire bond that has been enlarged because of excess lateral movement of the bonding tool or the holding fixture.

Snapback   In screen printing, the return of the screen to its normal plane after being deflected by the squeegee, after the squeegee has moved across the screen and substrate. *See also Screen Printing.*

Snap-Off Distance *See Breakaway.*

Snapstrate   A large substrate that contains multiple circuits and has been laser- or diamond-scribed so that it can be broken apart easily and without damage. *See also Scribe.*

Soak Time   The length of time, during the peak temperature of the firing cycle, in which a thick-film paste is held.

Softening Point   The point at which glasses lose their rigidity and become thermally mobile. For glass having a density of 2.5, this temperature corresponds to the log viscosity of 7.6 poises, per American Society for Testing and Materials (ASTM) Method C-338.

Soft Error   A memory state error that is caused by a process but leads to no permanent change to the physical condition of the device.

Soft-Flow Molding Materials   Thermosetting molding materials that can be molded at low pressures. *See also Transfer Molding.*

177

Soft Glass A type of glass that has a low softening point, approximately 450°C. Soft glass contains a high percentage of lead and is sometimes called *solder glass* because of its wettability to metal surfaces. It is used in hybrid packages for glass-to-metal seals, solder glasses, and glassivation, which is a thin-film passivation layer. *See also Passivation.*

Soft Solder (1) A lead-tin solder alloy with a low melting point, generally below 800°F (425°C) (2) An indium solder or other low-temperature-melting or soft soldering material.

Soft Substrate A substrate that is fabricated from a soft plastic material, as opposed to a hard ceramic or hard plastic substrate. The most common soft substrates are low-electrical-loss materials, such as fluorocarbons or polyethylenes, having a low dielectric constant and low dissipation factor. They are most often used for microwave and high-frequency electronic applications. *See also Low-Loss Dielectric, Substrate.*

Software Computer programs, procedures, rules, and associated documentation concerned with the operation of a data-processing system. *See also Hardware.*

Solar Array An array of solar cells or panels used to generate power from the sun and solar sources.

Solder An alloy with a relatively low melting point that is used to join two metals together, each having a higher melting point than the solder. Solders melt over a range of temperatures. The temperature at which the solder begins to melt is called the *solidus;* and the temperature at which the solder is completely molten is called the *liquidus. See also Phase Diagram, Vapor Phase Soldering, Wave Soldering.*

Solderability The ability of a metal to be easily wetted by solder and to form a strong bond with the solder.

Solder Acceptance *See Solderability, Wettability.*

Solder Balls Small spheres of solder that remain on the surface of a printed wiring board assembly after wave or reflow soldering. These balls must be removed, since they are sources of electrical shorts. *See also Reflux Soldering.*

Solder Bridge An electrical short circuit between two conductors, caused by solder that bridges the dielectric and electrically joins the two conductors.

Solder Bumps Solder balls that are bonded to contact areas or pads of semiconductor devices and that are subsequently used for face-down bonding. *See also Face Bonding.*

Solder Coat A layer of solder that is applied directly from a molten solder bath to a conductive pattern.

Solder Column *See Solder Post.*

Solder Connection An electrical and/or mechanical connection that employs solder for the joining of two or more metal parts. *See also Disturbed Solder Joint, Solder.*

Solder Contact A contact or terminal that has a cup into which a wire is inserted and soldered.

Solder Contact Terminal The location on a mother board to which a connector is soldered. *See also Mother Board.*

Solder Cream *See Solder Paste.*

Solder Cup The end portion of a terminal into which the conductor is inserted and subsequently soldered in place.

Solder Dam A screen-printable dielectric paste used for preventing molten solder from spreading onto conductors. *See also Screen Printing.*

Soldered Joint A metallic bond between a solder alloy and the surfaces to be bonded, such as leads, pads, terminals, and component surfaces. The three contributing factors that influence the quality of a soldered joint are the solder alloy selected, the flux, and the elimination of surface contaminants.

Solder Extraction A pulse vacuum technique for unsoldering component leads or wires from holes in a printed wiring board. It consists of a heated tip that melts the solder and a vacuum tube that sucks the molten solder from the hole. For planar-mounted components, a hot-air-jet method is used in which a stream of hot air melts the solder and the leads are removed without touching the solder joint. *Also called solder sucker.*

Solder Eye A solder-type terminal with a hole at one end into which a wire is inserted and subsequently soldered.

Solder Eyelet A contact with a hole into which a wire is inserted and mechanically attached before soldering.

Solder Fillet A meniscus-shaped configuration of solder around a component lead and the land to which it is soldered.

Solder Flux A material that prevents oxidation during heating and also transforms passive contaminated surfaces into clean, active, solderable surfaces. Flux should prevent oxidation during heating, lower interfacial surface tensions, be thermally stable, be easily displaced by molten solder, be noninjurious to components, and be easily removed with a mild solvent or aqueous solution. *See also listings under Flux.*

Solder Free Interconnections (SFI) Electrical interconnections made by meshing metal fibers together rather than by conventional soldering.

Solder Fusion A method that changes electroplated tin-lead on a circuit into a strong bond to the base

179

copper. The printed wiring board circuitry is first electroplated with a tin-lead alloy and subsequently heated to a temperature at which the tin-lead alloy reflows and fuses to the copper base material.

Solder Glasses A category of glasses used in package sealing because of their low melting point and the ease with which they wet and bond to metal and ceramic surfaces. *See also Soft Glass.*

Solder Immersion A test in which the metal leads or other parts of an electronic package are immersed in molten solder to determine their resistance to soldering temperatures.

Soldering A process of joining metals by fusion and solidification of an alloy having a melting point of less than 800°F and without melting the base metals.

Soldering Dissolution *See Dissolution Rate.*

Solder Land *See Pad.*

Solder Leaching The dissolving or alloying of the base metal to be soldered into the molten solder, usually caused by heating the base material to high temperatures. Gold often leaches into solder, so that thin gold coatings may completely dissolve, leaving no solder bond of the gold-coated parts.

Solderless Connection A connection formed between two metals by pressure only. No heat or solder is used.

Solderless Terminal A connection made by securing a wire inside a metal sleeve or similar device with a crimping tool. Solderless terminals vary in size and style and may be stronger mechanically than solder joints.

Solderless Wrapped Connection The connecting of a solid wire to a square, rectangular, or V-shaped terminal by tightly wrapping a solid conductor wire around the terminal with a special tool. *Also called wire-wrapped connection.*

Solder Leveling A process in which a heated gas is used to remove excess solder by melting and flowing the solder.

Solder Lugs Metal devices that readily accept a wire and are subsequently soldered. They are attached to printed circuit boards, termination strips and blocks, and electrical components.

Solder Mask *See Solder Resist.*

Solder Mask over Bare Copper (SMOBC) A solder mask coating applied directly over copper conductors on a printed board, as opposed to over solder-coated conductors. Advantages include cost and elimination of solder-bridging problems. *See also Solder Bridge, Tinning.*

Solder Oils Specially formulated liquids, for use in solder pots, wave-soldering equipment, and other devices, that cover the surface of the molten solder to prevent oxidation and the formation of dross. *Also called tinning and soldering oils. See also Dross.*

Solder Paste A homogeneous mixture composed of minute solder particles, solvent, flux, and other agents. It can be applied by dispensing equipment or screened onto the substrate. The deposited solder paste is normally heated to remove the nonsolder agents, so that the solder metal reflows into a pure solder joint. *Also called solder cream. See also Reflow Soldering.*

Solder Post A column of solder in an electronic assembly used for low-stress solder joints or as a thermal path. *Also called solder column. See also Self-Stretching Soldering Technology.*

Solder Preform A solder raw material formed into a specific shape, for control of solder joint thickness and convenience in fabrication or manufacturing. *See also Preform.*

Solder Projection An undesirable solder protrusion extending from a solder joint.

Solder Reflow *See Reflow Soldering.*

Solder Resist A coating material used to mask off and surface-insulate those areas where solder is not wanted, usually around component mounting holes on printed circuit

boards. Solder resists are available as epoxy-based pastes and in dry-film form. The thick solder mask film remains on the board to serve as an electrical and environmental barrier. *Also called solder mask.*

Solder Side (1) The secondary side of a single-sided printed board assembly. (2) The side of the board opposite the components. (3) The side of the board where the component leads are soldered to the circuit patterns.

Solder Sleeve A heat-shrinkable plastic tube containing a predetermined amount of solder and flux, used for environmentally resistant solder connections and shield terminations. The lead or wire is inserted into the tube, after which the connection is heated to the solder melting point. Simultaneously, the sleeve shrinks, resulting in a tightly soldered connection. *See also Heat Shrinkable.*

Solder Spheres Nearly perfect metal spheres engaging from approximately 0.01 inch minimum diameter ($\pm$0.002 inch) to a variety of larger sizes. Spheres are available as pure metals as well as alloy compositions. They have a wide range of applications, such as under surface-mount devices and microcomponents.

Solder Splatter Undesirable solder fragments that result from nonuniform soldering temperatures.

Solder Sucker *See Solder Extraction.*

Solder Webbing An undesirable solder bridge between two conductor patterns or circuits on a printed wiring board or substrate.

Solder Wetting The achieving of intimate contact between the liquid solder and the metal to be soldered, a contact required for a good, reliable solder joint. Inadequate solder wetting, which can occur for many reasons, results in a poor, improper, or incomplete solder joint. Solder wetting is both visually observable and measurable by surface tension measurement techniques.

Solder Wicking An undesirable flow of solder over the surface of each conductor of stranded wire, completely covering each strand and extending under the insulation. Wicking destroys the flexibility of the stranded wire, resulting in brittleness, and is usually caused by temperature differentials in the assembly.

Solid Logic Technology (SLT) The screening and firing of silver palladium conductors on an alumina substrate, as was practiced in the 1960s.

Solid Metal Mask A thin sheet of metal, normally less than 10 mils, with an etched pattern used in contact printing of film circuits.

Solid Phase Bond A bond between two parts in which no liquid phase occurred prior to or during the joining process.

Solid State Technology (SST) A technology in which components and circuits are made from semiconductor materials. *See also Integrated Circuit, Semiconductor.*

Solid Tantalum Chip A chip or leadless capacitor that contains a solid electrolyte, Ta_2O_5, instead of a liquid.

Solidus The highest temperature at which a metal or alloy is completely solid. *See also Liquidus.*

Solubility The extent that one material will dissolve in another.

Solute The substance dissolved in a solution. *See also Solvent.*

Solvent A liquid, usually organic or aqueous, that dissolves other substances. With respect to cleaning substances, there are three general categories of organic solvents: (1) Chlorinated hydrocarbons, such as methylene chloride, tetrachloroethylene, trichloroethylene, and methyl chloroform, which are nonpolar, nonflammable, and toxic, and are used in vapor degreasers. (2) Hydrocarbons, such as alcohols, ketones, and aliphatic compounds, which are highly flammable, polar, and medium toxic, and should never be used in vapor degreasers. (3) Fluorinated and fluorinated-chlorinated hydrocarbons, such as trichlorotrifluroethane, which are nonflammable, have low toxicity, are compatible with many materials, and are suitable for use in vapor degreasers. There are also a variety of azeotropic compositions, made up of solvents from the above three families, that are non-flammable and virtually nontoxic and can be used in vapor degreasers. *See also Solute.*

Solvent Cleaning The cleaning of electronic assemblies, especially circuit board assemblies, using organic solvents in the cleaning process. *See also Organic Solvents.*

Solvent Extract Resistivity Test A test for measuring the electrical resistivity of the solvent being used for cleaning electronic assemblies, especially circuit board assemblies. For example, the resistivity of deionized water is very high, but will be reduced after dissolving ionized impurities from the electronic assembly being cleaned. When the electronic assembly is thoroughly cleaned, no such contamination will be extracted in cleaning, and the resistivity of the deionized water will remain high even after the cleaning process. This test is therefore a means of measuring the cleanliness or contamination level of electronic assemblies. *See also Deionized Water.*

Solvent Resistant A material that is not affected by solvents and does not degrade when in contact with solvents.

Source A region of a semiconductor device from which majority carriers flow into the channel. *See also Channel, Dielectric Constant, Specific Inductive Capacity.*

Source Connector A connector that interconnects a light source, such as a light-emitting diode, to a fiber-optic cable.

Source Region Electromagnetic Pulse (SREMP) The electromagnetic pulse from a nuclear blast at or near the surface of the earth.

Spacing The linear distance between adjacent conductor edges.

Spade Connector A connector having a terminal with a slotted tongue and nearly square sides.

Spade Tongue Terminal A terminal with a slotted area to accommodate a screw or bolt that can be loosened or tightened for the removal or insertion of the terminal without removing the nut.

Spalling The chipping, fragmenting, or separation of a surface coating, or the cracking, breaking, or splintering of a material from heat.

Spark Gap A device with two electrodes that are separated by gas or air. After an electrical discharge, the insulation, air or gas, is self-restoring. It is used as a switching or protective device.

Specification A precise statement of a set of requirements of a material, system, or service and a procedure to determine whether the requirements have been met. *See also Standards.*

Specific Gravity The weight of a substance divided by the weight of the same volume of water at the same temperature. *See also Density.*

Specific Heat (1) The amount of heat required to raise the temperature of 1 gram of material 1°C. (2) The ratio of a material's thermal capacity to that of water at 15°C.

Specific Inductive Capacity *See Dielectric Constant.*

Specific Modulus Young's modulus divided by material density. *See also Young's modulus.*

Specific Strength Ultimate tensile strength divided by material density. *See also tensile strength.*

Specimen A sample of material or a device that is taken from a production lot to be used for testing. *See also Test Coupon.*

Spike A pulse with a very short time and a greater amplitude than the average pulse.

Spin Coating A coating process in which a predetermined amount of low-viscosity resin is dispensed on the top surface of a wafer or other substrate, after which the coated part is spun at high velocity to spread the coating onto the part in a uniform, thin layer. The coating is then heat-cured to its final protective state.

Spinel A single-crystal magnesium aluminum oxide substrate.

Spiral Flow Test A method for determining the flow properties of a thermosetting molding resin in which the resin flows along the path of a spiral cavity. This test is especially used for transfer-molding materials. The length of the material that flows into the cavity and its weight give a relative indication of the flow properties of the resin. *See also Transfer Molding.*

Splay The tendency of a drill bit to drill out-of-round, elliptically shaped holes that are not perpendicular to the drilling surface.

Split-Tip Electrode Electrode used in parallel-gap solder and parallel-gap weld operations. *See also Parallel-Gap Solder, Parallel-Gap Weld.*

Spray Coating The coating of electronic assemblies, such as circuit board assemblies, by spray techniques. *See also Dip Coating, Vapor Coating.*

Spray-Up Any of several techniques in which a spray gun is used as the processing tool. In reinforced plastics, for example, fibrous glass and resin can be simultaneously deposited in a mold or on a surface by spray techniques. In essence, roving is fed through a chopper and ejected into a resin stream, which is directed at the mold by spray systems.

Spring Finger Action A design feature of a contact permitting easy, stress-free spring action, which provides contact pressure or retention. This feature is used on printed circuit boards and socket contacts.

Sprue A channel through which molten plastic or metal flows in the molding of parts. These channels also connect the many cavities in a multicavity mold. *See also Gate, Runner.*

Sputter Cleaning The surface cleaning of a substrate by bombardment with argon or other types of gas ions. The object is to remove oxide films and thereby improve the adhesion of the subsequently deposited material layers to the surface of the substrate. *See also Back Sputtering, Plasma Etching.*

Sputtering A thin-film process for depositing material on a substrate. The substrate is placed in a vacuum chamber directly opposite a cathode made of the metal or dielectric to be sputtered or evaporated. The cathode is then bombarded with positive ions, and small particles of the cathode material deposit uniformly on the substrate. *See also Vacuum Deposition.*

Squeegee A straightedge rubber blade that pushes the paste or ink across the screen and through the mesh onto the substrate or workpiece. The squeegee is usually made of rubber having a durometer of 60 to 80. *See also Durometer, Screen Printing.*

Stabilization Bake Application of heat treatment to completed hybrid devices with the object of stabilizing circuit parameters. Baking is in

an air oven at approximately 150°C, for specified lengths of time. *See also Burn-In, Prebake.*

Stabilizers Chemicals used in plastics formulations to assist in maintaining physical and chemical properties during processing and service life. An example is an ultraviolet stabilizer, which is designed to absorb ultraviolet rays and prevent them from attacking the plastic.

Stacked Via *See Through Via.*

Stacking Sequence For a composite such as glass-epoxy or graphite-epoxy, the sequence of laying plies into the mold. Different stacking sequences have a great effect on off-axis mechanical properties. *See also Composite.*

Stainless Steel Screen *See Screen.*

Stair-Step Print In thick-film technology, a print that resembles the mesh of the screen along the edges of circuitry and resistors. This unwanted effect can be attributed to insufficient squeegee pressure or emulsion thickness. *See also Screen Printing, Squeegee.*

Staking With respect to adhesives, the bonding of components, wires, and other elements to a surface by applying small quantities of an adhesive material.

Standard Deviation A measure of variation of data from the average. It is equal to the root mean square of the individual deviations from the average.

Standard Epoxy Glass A mixture of approximately 60 percent epoxy resin and 40 percent fiber glass. *See also Laminated Plastics.*

Standardization The establishment of common terms, practices, criteria, processes, equipment, parts, and assemblies to achieve the maximum uniformity of items and optimum interchangeability of parts and components.

Standard Module An electronic module that is standardized for broad application in many electronic assemblies. *See also Module.*

Standards References that are used as a basis for comparison or calibration. *See also Specifications.*

Standoff (1) The gap between the substrate or printed wiring board and the bottom of a mounted component. A minimum gap is necessary in subsequent cleaning processes for the removal of flux and particulate that may fill this area and create contamination problems. (2) A metal post bonded to a conductor or substrate and raised above the surface of the circuit.

Standoff Insulator A post made of an insulating material that is used to support a wire above the surface of a structure to which it is mounted.

Staple Fiber *See Fiber Glass.*

Starved Area A part of a laminate or reinforced structure in which the resin has not completely wetted the fabric. *See also Resin-Starved Area.*

Starved Joint A joint that has an insufficient amount of adhesive to produce a satisfactory bond.

Static Burn-In *See Burn-In, Static.*

Static Charge An electrical charge that clings to the surface of an object. *See also Electrostatic Discharge.*

Static-Dissipation Material A material with a surface resistivity greater than 10^5 but not more than 10^9 ohms/square.

Static Electricity Electricity that is not moving. *See Electrostatic Charge.*

Static Electricity Discharge *See Electrostatic Discharge.*

Static Flex The installation of the flexible printed wiring assembly to a fixed position in a system.

Static-Safe Work Station (SSWS) A manufacturing work station having primary requirements of location in a static-controlled work area with static-dissipation benchtop work surface; conductive floor or floor mat (optional); approved grounding of benchtop to a common ground through a nominally 1 megohm resistor; wrist strap, to be worn in contact with bare skin, and grounding wire connected to the work surface through a nominally 1 megohm resistor; static-safe treated smock; and antistatic finger cots.

Secondary characteristics include the use of a static-safe chair with antistatic seatcover; caution and electrostatic discharge warning signs and labels; air ionizer equipment; electrostatic discharge protective bags, envelopes, trays, bins, tote boxes, kitting sets, foams, shunts, and transfer boxes; and grounded tools, such as solder iron-tip probes, hot plates, bonders, reflow soldering, and test equipment and microscopes.

Steady State A condition in a circuit in which the voltage, frequency, current, and so on, have settled down after the initial startup and remain constant thereafter.

Steatite A ceramic material consisting mainly of silicate of magnesium and used primarily in ceramic substrates.

Stencil A mask made of thin sheet material that is used to produce an image on the surface of a part. The image is formed by deposition of material through the holes in the stencil onto the surface of the part. *See also Screen Printing.*

Step and Repeat A process in which a circuit pattern is repeated several times to form multiple images in evenly spaced rows. The multiple images are subsequently used to print several substrates or printed wiring boards simultaneously. *See also Printing.*

Step Soldering A technique for making solder connections within an electronic package by using solder alloys with higher melting points initially and solders with lower melting points thereafter.

Stitch Bond (1) A point-to-point wiring connection system. Insulated wires are welded to hardware that has been inserted into holes in printed wiring boards. (2) A wire bond that is made by laying the wire on a bonding pad and scrubbing it with a capillary-type bonding tool, through thermocompression or ultrasonic means, to form the joint. *See also Solderless Wrapped Connection.*

Stop Plate A device attached to a crimping tool that correctly locates a terminal, splice, or contact in the tool prior to crimping. *Also called locator.*

Storage Control Element (SCE) A device that directs the transfer of data and interfaces among the channels, processor storage, and central processor. *Also called system-controlled element.*

Storage Hierarchy A combination of the memory elements and their controls. These two form the memory for the processor.

Storage Life *See Shelf Life.*

Storage Temperature The temperature at which a part or device, without any power applied, is stored.

Straight-Through Lead A component lead that extends through a hole of a printed wiring board and does not require bending.

Strain (1) The deformation resulting from a stress, as measured by the ratio of the change to the total value of the dimension in which the change occurred. (2) The unit change, due to force, in the size or shape of a body as compared with its original size or shape. Strain is nondimensional but is frequently expressed in inches per inch or centimeters per centimeter.

Strand A single, thin filament, such as a fine wire. A number of these may be twisted together to form a stranded wire. *See also Stranded Wire.*

Stranded Conductor *See Stranded Wire.*

Stranded Wire A wire conductor created by combining a number of smaller wires, normally twisted together, into a single conductor or cable. This stranding of fine wires results in a much more flexible wire or cable for a given current-carrying capacity.

Stratification The separation of nonvolatile components of a thick-film paste into layers during firing as a result of the differences in densities of the components. Stratification is more likely to occur with a paste containing heavy conductive particles and with prolonged or repeated firings.

Stray Capacitance *See Stray Circuit Element.*

Stray Circuit Element An unwanted electrical situation not intended in design but occurring in practice. For instance, close conductors may lead to stray capacitance, which in turn can result in circuit deviations or malfunctions. *See also Capacitance.*

Stress The unit force or component of force at a point in a body acting on a plane through the point. Mechanical stress is usually expressed in pounds per square inch. Stress on materials may also be other than mechanical, such as electrical or environmental.

Stress Corrosion A gradual deterioration of the mechanical properties of a material, usually accompanied by crack propagation, caused by the acceleration of applied stress. This normally occurs in a high humidity or other corrosive atmosphere.

Stress Free Pertaining to a material that has been partially annealed or stress-relieved so that its molecules are no longer in tension.

Stress Relaxation The time-dependent decrease in stress for a specimen constrained in a constant strain condition.

Stress-Relieve A heating process in which the molecules of a material are no longer in stress. This is accomplished by heat-cycling the material, such as reheating a film resistor, or by annealing metal or plastic parts.

Strip (1) To remove the insulation from a wire or cable. (2) To etch away unwanted surface material layers, usually by chemical processes.

Strip Contacts Contacts that are supplied in a continuous length and that can be used on automatic installation equipment.

Stripline (1) A microwave conductor on a substrate. (2) A transmission line configuration parallel and equivalent to two ground planes. *See also Microstrip, Strip Transmission Line.*

Strip Terminal Terminal material supplied in a continuous length and used on automatic crimping equipment. *Also called tape terminal.*

Strip Transmission Line A printed wiring transmission line configuration that consists of a printed conductor surrounded by dielectric material and shielded by copper ground planes on both sides. *Also called stripline.*

Structured Gate A frame on which several printed wiring assemblies are mounted and attached to a chassis by a hinge. It can be easily opened for access and servicing.

Stub (1) The wire that connects the inputs of a circuit to the main signal line. (2) A short length of transmission cable that is used as a branch to another transmission cable.

Stud (1) A metal post that is used for connecting wires. (2) A metal post that connects the top side of a printed wiring board to the bottom side or from one level of conductors to another in a multilayer substrate or printed wiring board.

Stylus A needle-shaped probe used to make electrical contact on a pad of a leadless device or a film circuit. *See also Probe.*

Subassembly Two or more parts that are combined into a single unit for later insertion into an electronic assembly. This is usually done for ease in assembly and disassembly and for servicing and maintainability. *See also Assembly, Module.*

Subcarrier Substrate A small substrate containing devices and circuits that is mounted on a larger substrate. *See also Daughter Card.*

Subminiaturization Any technique in which miniature components and circuits are used to reduce the size and volume of electronic packages.

Substrate The base material that provides a supporting surface for deposited or etched wiring patterns, for attachment of component parts, or for fabrication of a semiconductor device. *See also Ceramic-Based Microcircuits, Printed Wiring Substrate.*

Subsystem A part of a system that performs some specific function of the system. Examples are transmitters, receivers, or signal processors of a radar system. Subsystems can be removed intact and tested sepa-

189

rately. *See also Line-Replaceable Unit.*

Subtractive Process A process for obtaining conductive patterns by the selective removal of unwanted portions of a conductive foil. *See also Fully Additive Process, Semiadditive Process.*

Superconductor A conductor that offers little or no resistance to the flow of current. Superconductivity requires use of these superconductors at very low, often impractical operating temperatures.

Supported Hole A hole in a printed board whose wall surfaces have been reinforced by plating. *See also Unsupported Hole.*

Supporting Plane A planar structure that is added to a printed wiring board to provide additional mechanical strength, thermal conductivity, dielectric strength, electrical insulation, or mechanical stability. It can be added internally or externally.

Surface Channel A transfer channel created at the semiconductor-insulator interference.

Surface Creepage Voltage *See Creepage.*

Surface Diffusion The high-temperature injection of atoms into the surface layer of a semiconductor material to form the junctions. It is usually a gaseous diffusion process.

Surface Energy A measure of the ability of a material to wet or to be bonded, usually expressed as angle of wetting. Teflon, for instance, does not wet or bond easily and thus has a low surface energy. *See also Wetting.*

Surface Finish The geometric irregularities in the surface of a solid material. Expressed in microinch per inch.

Surface Insulation Resistance (SIR) Insulation resistance across the surface of an insulating material. *See also Insulation Resistance, Surface Resistivity.*

Surface Leakage The flow of current over the boundary surface of an insulator rather than through its volume. *See also Surface Resistivity.*

Surface-Mount Assembly (SMA) An electronic assembly that is manufactured with surface-mount components using surface-mount technology. *See Surface-Mount Technology, Type I SMA, Type II SMA, and Type III SMA.*

Surface-Mount Component (SMC) Any electrical or mechanical component that can be soldered to the surface of a substrate or printed wiring board. *See also Insertion-Mount Component.*

Surface-Mount Device *See Surface-Mount Component.*

Surface-Mount Technology (SMT) The technology of mounting flat leaded or leadless components and electronic packages on the surface of printed wiring boards, as opposed to insertion-mount technology. *See also Insertion-Mount Technology.*

Surface Preparation   The physical and/or chemical preparation of an adherend to render it suitable for adhesive joining. *See also Adherend.*

Surface Resistivity   The electrical resistance across the surface of an insulating or resistive material when measured between opposite edges of a unit square of any dimension. This value is expressed as ohms per square, and is similar to sheet resistivity. Surface resistivity is often lower, and hence more critical, than volume resistivity, because of surface films and surface contamination on electrical parts. This is especially true after exposure of parts to unclean environments, such as humid environments and most outdoor environments. *See also Sheet Resistivity, Volume Resistivity.*

Surface Tension   The tendency of the surface of a liquid to contract because of the intermolecular attraction of the molecules below the surface of the liquid.

Surface Texture   The smoothness or roughness of the surface of a material, often indicated in microinches or by a profilometer printout. Usually a surface-finish requirement is specified on the drawing in applications where the finish is critical. *See also Profile, Profilometer.*

Surfactant   (1) A substance that improves the wetting of the surface of a material. (2) A chemical that is added to a fluid to reduce its surface tension, thereby aiding in the penetration of small, confined areas.

Surge   A sudden and/or abrupt change in voltage and current in a circuit.

Surge-Protective Device   A device that reduces or diverts surges of currents in a circuit.

Surge-Withstand Capability   The ability of a circuit or an electrical system to resist damage to its components because of excessive voltages.

Susceptibility   A condition of a system that makes it easily affected by interferences, transients, and signals other than those to which the equipment was designed.

Swaged Leads   Component leads that have been deformed and secured on the noncomponent side of a printed circuit board for subsequent soldering to the board. *Also called swaging or swedging.*

Swaging   *See Swaged Leads.*

Swedging   *See Swaged Leads.*

Swim   The movement of components or circuit conductors in the *x-y* directions on a printed wiring board during reflow soldering or other exposure to heat. Conductor swimming is undesirable, since it

191

decreases the distance between conductors. The solution is to use higher-temperature boards or substrates, or to have better temperature control.

Swimming In thick-film technology, the lateral moving of a thick-film conductor pattern on molten-glass crossover patterns. *See also Swim, Thick-Film Technology.* See Fig. 16.

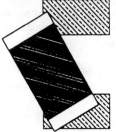

Figure 16: Swimming

Switching Noise A noise produced by an induced voltage at the circuit terminals. This induced voltage is created by the rapidly changing current caused by the switching of many devices.

Syntactic Foams Lightweight resin systems obtained by the incorporation of prefoamed or low-density fillers in the systems. Such fillers are usually hollow glass or plastic spheres. *See also Filler; Foam; Foam, In-Place.*

System A group of interconnected parts, devices, or subsystems designed for a specific end-product function, such as a radar system or a communications system.

System Control Element *See Storage Control Element.*

T

Tab *See Printed Contact.*

Tack The property of an adhesive that enables it to form a bond of measurable strength immediately after adhesive and adherend are brought into contact under low pressure. *See also Adherend.*

Tail of the Bond In wire bonding, the end of the wire that extends beyond the wire bond from the

heel. The tail is removed later, after the wire bonding is completed. *See also Wire Bonding.*

Tail Pull In wire bonding, the removing of the tails from the wire bonds. *See also Tail of the Bond.*

Tan Delta *See Dissipation Factor.*

Tantalum Capacitor An electrolytic capacitor that uses tantalum foil or a sintered slug of tantalum as the anode. The tantalum oxidizes and forms the dielectric barrier.

Tape and Reel A technique for transporting components that uses a tape with cavities shaped and sized for a family of products that are wound around a reel.

Tape-Automated Bonding (TAB) A process in which precisely etched leads, which are supported on a flexible tape or plastic carrier, are automatically positioned over the bonding pads on a chip. A heated pressure head is then lowered over the assembly, thereby simultaneously thermocompression-bonding, or gang-bonding, the leads to all the pads on the chip. *See also Gang Bonding.*

Tape Carrier Package (TCP) A packaging technique that uses tape-automated bonding (TAB) technology for the internal integrated circuit connections and a postmolded package between the chip and the substrate. *See also Tape-Automated Bonding.*

Tape Carrier Ring (TCR) A package similar to a tape carrier package, but with a plastic ring to support the outer leads during test, burn-in, and shipment. *See also Tape Carrier Package.*

Tape Casting The fabrication of a green tape ceramic, as opposed to a dry-pressing or pressed-powder technique. In tape casting, casting powder is added to organic liquid and dispersing agents, after which the mixture is blended in a roller mill to form a liquid slurry. This slurry is poured onto a casting table, and the liquids are evaporated off. The result is a flexible green tape that can be pinched, drilled, formed, and otherwise handled prior to final firing into the rigid ceramic part. *See also Green Tape, Roller Mill, Slurry.*

Taped Components Components that are attached to continuous tape by a sticky substance to facilitate automatic inspection, lead forming, testing, and subsequent transfer to printed boards or substrate by automatic equipment.

Tape Terminal *See Strip Terminal.*

Tape Wrap The application of thin insulating film or tape by wrapping the film or tape around the wire or wire bundle.

T Dimension *See G Dimension.*

Tear Strength A measurement of the amount of force needed to tear a solid material or tape that has been nicked on one edge and then subjected to a pulling stress. Measured in pounds per inch of material width.

Teflon *See Tetrafluoroethylene.*

Temperature Aging The stressing of a material, component, or system at an elevated temperature for an extended period of time. *See also Accelerated Aging.*

Temperature Coefficient The ratio of the change in a parameter to the change in temperature. The change in the parameter may or may not be normalized to the initial value of the parameter. This ratio is usually the average value of the total temperature change. The specific term should be *temperature coefficient of parameter*.

Temperature Coefficient of Capacitance (TCC) The amount of change in capacitance of a capacitor, or any electronic device, with respect to temperature. Measured in parts per million per degree Celsius (ppm/°C) over a specified temperature range.

Temperature Coefficient of Linear Expansion The amount of change in any linear dimension of a solid arising from a change in temperature. Measured in microinches per inch per degree Celsius. *Also called coefficient of thermal expansion (CTE)*.

Temperature Coefficient of Resistance (TCR) The maximum change in resistance of a resistor or any resistor material per unit change in temperature, usually expressed in parts per million per degree Celsius (ppm/°C) and specified over a certain temperature range. Sometimes expressed as an average change over a certain temperature in ppm/°C. The temperature is that of the resistor, not the ambient temperature.

Temperature Cycling An environmental test in which a material, component, or system is subjected to temperature changes, usually from a low temperature (−55°C) to a high temperature (+125°C) over a period of time and for a specific number of cycles (usually 10, 25, or 100 cycles). *See also Accelerated Aging, Accelerated Stress Test.*

Temperature Excursion The temperature range, from the lowest temperature to the highest temperature, that a component, material, or system experiences.

Temperature Profile The description of the temperature that a selected point traverses as it passes through a programmed heating cycle.

Tensile Strength (1) The maximum tensile stress that a material is capable of sustaining. Tensile strength is calculated from the maximum load during a tension test carried to rupture and the original cross-sectional area of the specimen. (2) The maximum amount of axial load required to pull a wire from a crimped barrel of a terminal.

Tented Via A blind or through via that has the exposed surface on the primary, or secondary, or both sides of a packaging and interconnecting structure fully covered by a masking material, such as a dry-film polymer covering, solder mask, or preimpregnated glass cloth (prepreg), in order to prevent hole access by process solutions, solder, or contamination. *See also Through Via, Via, Via Hole.*

Terminal (1) A metallic pin or point to which an electrical connection can be made. (2) A device designed to handle one or more connectors, either insulated or bare, that are to be affixed to a board, bus, stud, or chassis, in order to establish an electrical connection. Some of the most common types of terminals are solder contact, clip, package lead, solderless wrap, spade, ring, flag, blade, flanged, and offset.

Terminal Area *See Pad.*

Terminal Block A block of insulating material on which many terminal connectors have been mounted for making electrical connections.

Terminal Board (1) A board made of an insulating material that has a single or double row of termination pads for making electrical connections to a mating connector. (2) An insulating board equipped with terminals for making wire connections.

Terminal Clearance Hole *See Access Hole.*

Terminal Hole *See Component Hole.*

Terminal Pad *See Pad.*

Termination *See End Termination.*
Ternary Pertaining to materials that have three elements. *See also Polynary.*

Testability The degree or ease to which an electronic circuit or package can be electrically tested.

Test Coupon A sample or test pattern, usually made as an integral part of the printed board, on which electrical, environmental, and microsectioning tests may be made to evaluate board design or process control without destroying the basic board. *Also called test pattern. See also Specimen.*

Test Pattern A circuit configuration on the semiconductor wafer that provides test sites for monitoring fabrication processes.

Tetrafluoroethylene (TFE) The original and most common form of DuPont Teflon. Now known as TFE and made by several suppliers.

Thermal Relating to heating and cooling and the control thereof.

Thermal Barrier Any resistance to the flow of heat. *Also called thermal impedance.*

Thermal Coefficient of Expansion *See Coefficient of Thermal Expansion.*

Thermal Conduction Module (TCM) A module containing 100 chips or more that is cooled by thermal conduction devices in contact with the chips.

Thermal Conductivity (1) The ability of a material to conduct heat through itself. (2) The physical constant for the quantity of heat that passes through a unit cube of a material in a unit of time when the differences in temperature of the two faces is 1°C. The rate at which a material conducts heat is expressed in calories per square centimeter per centimeter per cross section per degree Celsius. Thermal conductivity varies with temperature. Average values of thermal conductivity for some common electronic packaging materials are:

Teflon	0.0006
Unfilled epoxy	0.0006
Filled epoxy	0.001 – 0.05
Glass	0.003
Kovar	0.04
Aluminum	0.48
Copper	0.92
Diamond	1.9

Thermal Cycling A test technique of raising and lowering temperatures, used to induce a stress on electrical components or assemblies. *See also Accelerated Aging, Accelerated Stress Test.*

Thermal Design The optimized design of an electronic package or system based on a study of the flow of heat and heat paths from each heat-dissipating component in a circuit to an external heat sink.

Thermal Drift The change in heat dissipation from nominal values for components and circuit elements as a result of changes in temperature and other specified variables.

Thermal Drop The difference in temperature across a boundary or across a material.

Thermal Expansion The expansion of a material or part as a function of temperature. *See also Coefficient of Thermal Expansion.*

Thermal Expansion Coefficient *See Coefficient of Thermal Expansion.*

Thermal Expansion Mismatch The differences in coefficients of thermal expansion of two materials that are joined together, such as an electronic package soldered to a printed wiring board. This mismatch may be critical for solder joint reliability. *See also Coefficient of Thermal Expansion.*

Thermal Gradient The variation in temperature across the surface of or through a material being heated.

Thermal Grease A thermally conducting grease, usually a filled silicon, that can be placed under a heat-generating component or part to remove heat from that component or part. *See also Thermal Pad.*

Thermal Management All aspects of the control of heating in electronic packages and systems.

Thermal Network A model of a total system that is broken down into subsystems, each displaying its thermal property and connected to the other members of the subsystem so as not to distort the total thermal property of the system.

Thermal Noise An undesirable operating condition caused by thermal motion of charged particles within an electronic component. Such noise can lead to degradation of component or system performance.

Thermal Pad A thermally conductive pad, usually a filled soft plastic material, that can be mounted under a heat-generating component or part to remove heat from that component or part. *See also Thermal Grease.*

Thermal Relief The crosshatching of a ground or voltage plane that minimizes blistering or warping during soldering operations.

Thermal Resistance The resistance offered by a material or medium to the flow of thermal energy through the medium.

Thermal Resistance, Case-to-Ambient The thermal resistance (steady-state) from the semiconductor package to the ambient.

Thermal Resistance, Junction-to-Ambient The thermal resistance (steady-state) from the virtual junction of the semiconductor device to the ambient. *See also Virtual Junction.*

Thermal Resistance, Junction-to-Case The thermal resistance (steady-state) from the virtual junction of the semiconductor device to a stated location on the case. *See also Virtual Junction.*

Thermal Resistance, Steady-State The temperature difference between two specified points or regions divided by the power dissipation under conditions of thermal equilibrium. Examples are case to ambient, junction to ambient, junction to case, and junction to lead.

Thermal Shock An environmental test in which devices or equipment are rapidly subjected alternately to high and low temperatures, as in accelerated testing, in an attempt to cause early failures. *See also Accelerated Aging, Accelerated Stress Test.*

Thermal Stability The resistance of a material to changes in physical, electrical, and chemical properties caused by heat from elevated temperatures.

Thermal Stress Cracking Crazing and cracking of some thermoplastic resins as a result of overexposure to elevated temperatures.

Thermal Via A hole in a circuit board assembly that is filled with some thermally conductive material and that transfers heat from heated components to some heat sink. *See also Through Via.* See Fig. 17.

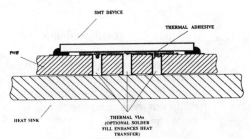

SMT DEVICE

THERMAL ADHESIVE

PWB

HEAT SINK

THERMAL VIAs
(OPTIONAL SOLDER
FILL ENHANCES HEAT
TRANSFER)

Figure 17: Thermal Via

197

Thermocompression Bonding The joining of two materials by interdiffusion across the boundary between them by the application of heat and pressure.

Thermoforming The creation of a form from a flat sheet by combinations of heat and pressure that first soften the sheet and then form the sheet into some three-dimensional shape. This is one of the simplest and most economical plastic-forming processes. There are numerous variations for both thermoplastics and thermosetting plastics. See Fig. 18.

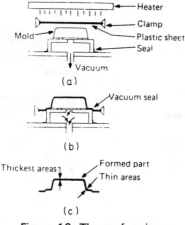

(a)

(b)

(c)

Figure 18: Thermoforming

Thermogravimetric Analysis (TGA) A method that measures the change in weight of a material as a function of increasing the temperature of the material.

Thermomechanical Analysis (TMA) A method that measures the linear expansion or contraction of a material as the temperature of the material is increased or decreased.

Thermoplastic A plastic that is set into its final shape by forcing the melted base polymer into a cooled mold or through a die, after which it is cooled. The hardened plastic can be remelted and reprocessed several times. *Also known as linear or branched polymer.*

Thermoset A plastic that is cured, set, or hardened, usually by heating, into a permanent shape. The polymerization is an irreversible reaction known as *crosslinking.* Once set, thermosetting plastics cannot be remelted, although most will soften with the application of heat.

Thermosonic Bonding A bonding process that combines thermocompression and ultrasonic bonding. A gold wire is bonded to a die or pad on a substrate by applying ultrasonic power to the capillary of a thermocompression bonder, resulting in a bond or weld at a temperature of approximately 150°C.

Thick Film A layer of conductive, resistive, or dielectric ink or paste screen-printed onto a substrate, which is then heated to form conductors, resistors, and capacitors. Heating is known as *firing* for cermet thick films or *curing* for polymer thick films. Layers are typically about 0.001 inch thick. *See also Cermet Thick Film, Polymer Thick Film.*

Thick-Film Circuit A microcircuit, including passive devices such as resistors and capacitors that are screen-printed onto a ceramic substrate and fired.

Thick-Film Dielectric A screen-printable paste that contains finely ground insulating materials such as glass powders or ceramic powders.

Thick-Film Hybrid Circuit A thick-film network on a substrate to which chip devices or thick-film devices have been added to form a hybrid microcircuit that in turn performs an electronic function.

Thick-Film Network A network of screen-printed resistors and capacitors that are interconnected by screen-printed conductors on a ceramic substrate and subsequently fired.

Thick-Film Resistor, Conductor, and Dielectric Compositions Screen-printable pastes that contain metal oxides, metals, or glass powders, respectively.

Thick-Film Technology A technology in which electronic circuits and networks are formed by screen-printing conductive, resistive, and dielectric layers on a ceramic substrate and firing. Conductive layers are approximately 0.0003 to 0.001 inch thick; dielectric, 0.001 to 0.002 inch thick; and resistors, 0.0004 to 0.0008 inch thick.

Thief *See Plating Thief.*

Thin Film A layer of conductive, resistive, and dielectric material sputtered or evaporated onto a substrate in a vacuum to form conductors, resistors, and capacitors. Layers are typically less than 1000 angstroms thick (1 angstrom is approximately 4×10^{-9} inch or 1×10^{-8} cm).

Thin-Film Capacitor A capacitor that is made by the evaporation or sputtering of conductive and dielectric materials in the form of layers.

Thin-Film Circuit A circuit in which active or passive devices and conductors are produced as films on a substrate by the evaporation of dielectric, resistive, and conductive materials.

Thin-Film Hybrid Circuit A thin-film network on a substrate to which chip devices or thin-film devices have been added to form a hybrid microcircuit that in turn performs an electronic function.

Thin-Film Integrated Circuit A microcircuit in which passive devices are produced as films by evaporation or sputtering techniques on a substrate.

Thin-Film Network A network of resistors and/or capacitors that are interconnected by conductors on a substrate and vacuum-deposited by evaporation or sputtering techniques.

Thin-Film Packaging Electronic packages whose substrates are fabricat-

199

ed using thin-film conductors and insulators.

Thin-Film Technology A technology in which electronic circuits and networks are formed by vacuum evaporation, sputtering, or other deposition techniques on a substrate. *See also Thick-Film Technology.*

Thinner A volatile liquid added to an adhesive or paint to reduce the viscosity of the material or compound. *See also Diluent.*

Thin Small Outline Package (TSOP) A thin, rectangular plastic package, usually with leads only on the ends, that is used for memory-integrated chips.

Thixotropic Materials that are gellike at rest but fluid when agitated. Most resins can be made thixotropic by the addition of large-surface-area fillers such as flocculated silica powder. *See also Filler.*

Three-Dimensional (3-D) Circuit Boards *See Molded Circuit Boards.*

Three-Layer Tape In tape-automated bonding (TAB), a tape that contains two interconnected metallized layers separated by a dielectric layer. *See also Tape-Automated Bonding.*

Through Connection An electrical connection between conductors on opposite sides of a substrate or base material. *See also Interlayer Connection, Buried Via, Through Via.*

Through Hole *See Plated Through Hole. See also Through Via, Via, Via Hole.*

Through-Hole Component A leaded component designed for mounting on a printed board by inserting the leads through the holes and subsequently soldering them to circuitry on the board.

Through-Hole Mounting The electrical connection of components to a conductive pattern by the use of component holes.

Through Via A fixed via that extends through all the metallized layers of a thick-film multilayer substrate. It can be also used for electrical grounding or thermal dissipation. *Also called stacked via or thermal via.*

Throwing Power The ability of an electroplating bath to deposit metal particles in recess areas, holes, and low-current-density areas without regard to the thickness of the part.

Thyristor A bistable device that comprises three or more junctions and can be switched from the off state to the on state or vice versa.

Time Constant The time required for a signal to rise to 63.2 percent or delay to 36.8 percent of maximum.

Time Domain Reflection A rise and fall in current and voltage at the point of a break in a transmission line, creating a disturbance or reflection, that travels in the opposite direction and causes line noises and signal distortions.

Tin (Sn) A widely used metal for plating brass, copper, and steel terminals. It has excellent electrical and thermal conductivity and is used on component leads to increase solderability and to reduce galvanic corrosion when in contact with aluminum. It also is used for alloying with other metals to improve mechanical and physical properties. Examples are tin-nickel (65 percent tin, 35 percent nickel) and tin-lead (62 percent tin, 38 percent lead).

Tin-Lead (Sn-Pb) An alloy, usually 62 percent tin and 38 percent lead, used in most soldering applications because of its low melting point. *See also Solder, Soldering.*

Tinning The process of coating metallic surfaces with a thin layer of tin or solder.

Tinning and Soldering Oils *See Solder Oils.*

Tip In wire bonding, the part of the bonding tool that applies pressure to deform the wire and to form the bond. *See also Wire Bonding.*

TO Can *See TO Package.*

Toe *See Tail of the Bond.*

Tombstoning The raising up of small chips from the substrate during solder reflow as a result of surface tension and unbalanced forces of solder wetting. This can result in

an open circuit at one end of the chip or an open wire bond. *Also called drawbridging and Manhattan effect.* See Fig. 19.

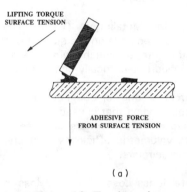

(a)

Figure 19: Tombstoning

TO Package A generic term for a transistor outline package, established by JEDEC as an industry standard. The package consists of a cylindrical shaped metal can containing an integrated circuit chip. The chip is mounted on a base and interconnected to terminals with feed through glass to metal seals. A metal cover, in the form of a top hat, is placed over the integrated circuit and hermetically sealed to the metal base. *See also Transistor.*

Top-Hat Resistor A thick-film resistor with a top-hat shape. *See also Thick-Film Resistor.*

Topography In thick-film technology, the condition of a fired surface with respect to valleys and bumps in the surface profile. *See also Profile, Profilometer.*

201

Topology A surface layout design, study, and characterization of the elements of an integrated circuit. It is used in creating the artwork for mask preparation.

Top-Side Metallurgy (TSM) The metallized pads on the surface of a substrate to which chips are attached or electrically connected.

Toroid A ring- or doughnut-shaped coil winding that is used for inductors and transformers because of its volumetric efficiency in hybrid microcircuits.

Total Mass Loss The total loss in mass of a material resulting from exposure of the material to thermal-vacuum conditions, such as exist in outer space. This loss consists largely of the evaporation of volatile condensible materials. *See also Volatile Condensible Materials.*

Touch-Up The elimination of defects in a product, such as photographic artwork or a printed board, prior to etching, usually achieved by hand operation.

Tow An untwisted bundle of continuous fibers. Commonly used in reference to synthetic fibers, particularly carbon and graphite but also glass and aramid. A tow designated as 12K has 12,000 filaments.

Track (1) A path of deteriorated material on the surface of a dielectric. *See also Tracking.* (2) A conductor on a substrate.

Tracking The conductive carbon path formed on the surface of a plastic during electrical arcing. *See also Arc Resistance.*

Track Resistance The resistance of organic materials, such as plastics, to the formation of a carbon track on the surface of the material by high-voltage arcing. *Also called arc resistance.* Precisely defined, *arcing* is the electrical condition and *tracking* is the carbon path resulting from deterioration of the material by the arcing condition. *See also Arc Resistance.*

Transfer Bump Tape-Automated Bonding Tape-automated bonding that uses discrete bumps between the die lands and carrier tape to facilitate inner lead bonding. *See also Tape-Automated Bonding.*

Transfer Lamination *See Translam Transfer Lamination.*

Transfer Molding A method of molding thermosetting materials in which the plastic is first softened by heat and pressure in a transfer chamber, then forced by relatively low pressure through suitable sprues, runners, and gates into a closed mold for final curing. Electronic assemblies can be safely molded by this process because of the relatively low molding pressures. *See also Compression Molding, Injection Molding.*

Transient A sudden change in conditions in a system, such as a surge on a signal or power line, that lasts for a short period of time. The change may result in insulation and/or component breakdown and failure.

Transient Control Level The highest peak voltage that a system can take without causing damage to the circuit or system.

Transient Mismatch A difference in thermal conductivity of two materials or parts in a system. Because of this difference, a thermal lag occurs until equilibrium is reached. *See also Thermal Gradient, Thermal Network.*

Transient Radiation Effects on Electronics (TREE) The damage to the electronics of a system caused by photons or subatomic particles. *See also Photon.*

Transistance The electrical property that effects voltages or currents so as to accomplish gain or switching action. Examples of transistance occur in transistors, diodes, voltage-controlled rectifiers, and electron tubes.

Transistor An active semiconductor device having three or more electrodes. It is usually made of germanium or silicon and is used as an amplifier, a rectifier detector, or a switch. *See also TO Package.*

Transistor Outline Package *See TO Package.*

Transistor-Transistor Logic (TTL) (1) A widely used, inexpensive form of semiconductor logic with relatively high speed and medium power dissipation. Its basic logic element is a multiple-emitter transistor. (2) A design technique that uses bipolar transistors. Logic circuits designed by this technique have a static threshold of 1.5 volts.

Translam Transfer Lamination An additive plating and laminating process for producing fine-line printed circuit boards. Rapid impingement speed plating (RISP) is used as the plating process. Each layer is plated on a stainless steel plate, covered with B-stage dielectric material, and transferred to a board stack-up for final lamination. *See also Rapid Impingement Speed Plating.*

Transmission Line (1) A conductor or a series of conductors used for carrying electrical energy from a source to a load. (2) A conductor that is inductively and capacitively coupled to a nearby return path to form a uniform distributed network. It is used for transmitting pulse signals.

Transmission Loss The amount of power lost by a signal as it is transmitted from one point to another. The loss is usually expressed in decibels. *See also Decibel.*

Transverse Isotropy Having essentially identical mechanical properties in two directions, but not the third.

Transverse Properties Properties perpendicular to the axial (x,1 or 1,1) direction. May be designated as y or z, 2 or 3 directions.

Treating Bath The catalyzed resin mixture that is used to resin-coat the glass fabric in making a laminated plastic. The coated fabric is then partially or fully cured in a heating oven. *See also Laminated Plastics.*

Treeing (1) A dendritic type of plating growth that extends onto the surface adjacent to the edge of a conductive pattern. It is normally caused by excessive plating current. *See also Dendritic Growth.* (2) The formation of structures, whose appearance resembles tree roots, when certain plastic materials are subjected to excessively high voltage gradients. These structures are internal to the plastic part, and result from the high energy stress produced by the high voltage condition. The presence of such internal fissures makes the plastic material unsuitable for further use in high-voltage applications. *See also Corona, Tracking.*

Tribology The science of lubrication.

Trifunctional Epoxy An epoxy resin having three epoxide reactant units per epoxy molecule. The increased number of reactant units per molecule results in a cured epoxy that has greater thermal stability than that obtained with difunctional epoxies. *See also Difunctional Epoxy, Multifunctional Epoxy.*

Trimming *See Functional Trimming.*

Tuning Adjusting a variable resistor, capacitor, or other component so as to change system operation.

Turbulent Flow A type of flow in which any fluid particle may move in any direction with respect to any other particle. *See also Laminar Flow.*

Twill Weave A basic weave characterized by a diagonal rib or twill line. Each end floats over at least two consecutive picks, creating a greater number of yarns per unit area than a plain weave, with relatively little loss of fabric stability.

Twist The spiral turns about the axis of a textile strand per unit length. Expressed as turns per inch.

Twisted Pair Two insulated wires that are twisted around each other without a common covering.

Two-Layer Tape A dielectric tape, such as Kapton, containing a metalized layer on one side of the tape that is etched to form the lead-frame used in tape-automated bonding (TAB). *See also Tape-Automated Bonding.*

Type I SMA A surface-mount assembly (SMA) composed entirely of surface-mount components (SMCs), with all the connections made by solder-reflow processes. Type I includes both single- and double-sided boards.

Type II SMA A surface-mount assembly (SMA) that has a mixture of surface-mount components (SMCs) and through-hole components on the same side of the printed board, with the connections made by solder-reflow and flow-soldering processes. Type II can be a double-sided printed wiring board, but one side must have a mixture of components.

Type III SMA   A surface-mount assembly (SMA) that has insertion-mount components (IMCs) on the primary side of the printed board

and surface-mount components (SMCs) adhesively bonded to the secondary side of the board. All the solder connections are made simultaneously by solder-reflow processes, with the bonded surface-mount components on the secondary side. *See also Insertion-Mount Component.*

U

Ultra-Fine-Pitch Technology (UFPT)   Any interconnection process in which the pitch, or distance between connection centers, is less than 0.2 inch. *See also Fine-Pitch Technology.*

Ultra High Frequency (UHF)   In the radio frequency spectrum, the band extending from 300 to 3000 mHz.

Ultra-Large-Scale Integration (ULSI)   A semiconductor chip that contains a minimum of 100,000,000 transistors.

Ultrasonic Bonding   (1) A joining process in which two metals are bonded together by applying pressure plus an ultrasonically induced scrubbing action to form a molecular bond. (2) The bonding of plastics by vibratory mechanical pressure at ultrasonic frequencies, which results in the melting of the plastics as they are joined.

Ultrasonic Cleaning   A cleaning method that utilizes the cavitation of a chemical solvent. The cavitation is derived from ultrasonic energy from a high-frequency generator and a transducer.

Ultraviolet Cure   A process in which crosslinking of certain polymer materials, such as epoxy adhesives and film emulsions, is activated by exposure to ultraviolet radiation.

Ultraviolet (UV) Radiation   Electromagnetic radiation in the ultraviolet region of the electromagnetic spectrum, having wavelengths between 200 and 4000 angstroms. *See also Electromagnetic Radiation.*

Ultraviolet Stabilizer   An additive mixed into a plastic formulation for the purpose of improving resistance of the plastic to ultraviolet radiation.

Uncured The state of a molded article that has not been adequately polymerized or hardened in the molding process, usually due to inadequate temperature-time-pressure control.

Underbond Insufficient deformation of the wire by the bonding tool in the wire-bonding operation.

Undercut The reduction of the cross-section of a material caused by etching action spreading beneath the edge of the photoresist or other masking film. *Also called over-etching.*

Underglaze (1) In thick-film technology, the application of a glass or ceramic glaze material to a substrate, at the point where the resistors are to be placed, prior to the screening of resistors. (2) A low-

gloss surface caused by insufficient frit material. *See also Frit.*

Universal Pattern A circuit board pattern or patterns that will accept standard-size packages and configurations, such as dual in-line packages.

Unsupported Hole A hole in a printed board that does not contain plating or other type of conductive reinforcement. *See also Supported Hole.*

Urea A thermosetting resin characterized by good chemical and electrical properties. It is known for its hard, scratch-resistant finish and high arc resistance as well as resistance to heat.

UV Stabilizer *See Ultraviolet Stabilizer.*

V

Vacuum Bag Molding A process for molding reinforced plastics in which a sheet of flexible transparent material is placed over the lay-up on the mold and sealed. A vacuum is applied between the sheet and the lay-up. The entrapped air is mechanically worked out of the lay-up and removed by the vacuum, and the part is cured.

Vacuum Bake A process in which unsealed packages and printed boards are placed in a vacuum chamber and heated to temperatures up to 150°C for periods of time, from several hours to several days, while constantly being pumped down to vacuum pressures of 2 to 5 mm of mercury to remove moisture and other contaminants from the packages and boards.

Vacuum Deposition The deposition of thin metal or dielectric films on a substrate by the evaporation of the materials in a vacuum chamber. *See also Sputtering, Vacuum Evaporation, Vapor Coating.*

Vacuum Evaporation The vaporizing by heating of metal and dielectric materials in a vacuum at reduced pressures. These materials are deposited as thin films on a substrate, and subsequently used as microcircuits, resistors, capacitors, and semiconductor devices. *See also Thin-Film Technology.*

Vacuum Injection Molding A molding process using a male and female mold in which reinforcements are placed in the mold, a vacuum is applied, and a curing liquid resin is introduced at room temperature to saturate the reinforcement.

Vacuum Pickup A tool used for picking up chip devices. The pickup end consists of a pencil-shaped tube with a nonscratching tip, while the other end is connected to a low-vacuum source.

Vapor Coating The coating of electronic assemblies by vapor or vacuum techniques, as in Parylene coating. *See also Dip Coating, Parylene, Spray Coating.*

Vapor Degreasing The removal of contaminants, oils, and greases from electronic assemblies using hot solvent vapors. The high solvent power of hot vapors creates a highly effective cleaning technique.

See also Cleaning, Degreasing, Solvents. See Fig. 20.

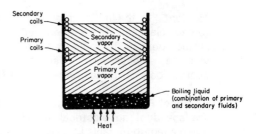

Figure 20: Vapor Degreasing

Vapor Deposition *See Vacuum Deposition.*

Vapor Phase Reflow *See Vapor Phase Soldering.*

Vapor Phase Soldering A method for soldering many joints simultaneously. The process is performed in a heated chamber containing boiling fluorinated hydrocarbons with two separate temperature zones — for example, Freon TF, which has a vapor phase zone of 185–215°F, and FC-5312, which as a boiling point of 419°F. Parts and assemblies are heated by condensation as the cold parts are introduced into the hot vapor. *See also Reflow Soldering.*

Varnish A liquid resin that is used to coat electrical components and impregnate electrical coils to provide electrical, mechanical, and environmental protection. *See also Impregnate.*

Vehicle A liquid that is added to thick-film paste or other viscous or paste materials to adjust the viscosity. *See also Diluent, Thinner.*

Veil Coat In the molding of reinforced plastics, a resin-enriched surface next to the mold that provides a smooth coat over the coarse fibers of the reinforcement.

Velocity of Propagation The ratio of the speed of an electrical signal down a length of cable to its speed in free space. It is the reciprocal of the square root of the dielectric constant of the cable insulation and is expressed as a percentage. *See also Propagation Time.*

Vent A small opening placed in a mold for allowing air to exit the mold as the plastic molding material enters. This eliminates air holes, voids, or bubbles in the finally molded plastic part.

Very High Frequency (VHF) In the radio frequency spectrum, the band extending from 30 to 300 mHz.

Very High Speed Integrated Circuit (VHSIC) An integrated circuit meeting the following phase goals of speed and density established by the U.S. Department of Defense: (1) minimum product of operating frequency and equivalent gate density at a Phase 1 value of 5 x 10^{11} Hz gates/cm^2 and a Phase 2 value of 1 x 10 Hz gates/cm2; and (2) minimum clock frequency at a Phase 1 value of 25 mHz and a Phase 2 value of 100 mHz.

Very Large Scale Integration (VLSI) An integrated circuit chip that contains a minimum of 10,000 transistors.

Very Low Frequency (VLF) In the radio frequency spectrum, the band extending from 10 to 30 kHz.

Vesical A blister formed as the result of vesication. *See also Vesication.*

Vesication The formation of blisters at the interface between a semipermeable polymer film coating and another material, caused by an osmotic effect from the interaction of water-soluble matter with moisture. *See also Mealing.*

Via (1) An opening in a dielectric layer that is filled with a conductive paste to form an interconnection between multilayers of a thick-film hybrid circuit. (2) An electrically conductive path that passes through an insulating material and connects conducting layers in two or more planes. *See also Blind Via, Buried Via, Fixed Via, Interstitial Via, Programmable Via, Stacked Via, Thermal Via, Through Via.*

Via Hole A plated through hole whose sole purpose is to interconnect two or more internal layers on a multilayer printed board, not to insert a component lead.

Vicat Softening Temperature The temperature at which a specified needle point will penetrate a material under preset test conditions. An example is the temperature at which a flat-ended needle having a 1 mm^2 circular cross section will penetrate a thermoplastic specimen at a depth of 1 mm under a specified load at a predetermined, uniform rate of temperature rise.

Video Pertaining to the function of detecting, transmitting, or processing signals having frequencies that lie within the range sensed by the human eye. These frequencies range from 100 kHz to several mHz and are primarily used for television.

Virtual Junction The theoretical point or region in a simplified model of the thermal and electrical behavior of a semiconductor device at or in which all the power dissipation within the device is assumed to occur.

Viscoelastic A characteristic mechanical behavior of some materials that is a combination of viscous and elastic behaviors.

Viscometer An instrument capable of measuring the viscosity of fluids and pastes.

Viscosity A measure of the resistance of a fluid to flow in a simple liquid or Newtonian fluid. The unit of measurement is the Pascal second (Pa.s). Frequently the centipoise (cp), which is one milliPascal second (m Pa.s), is used as the viscosity unit. For simple liquids, the viscosity is constant at all shear rates. In thick-film paste, a non-Newtonian liquid, the viscosity will vary depending on the shear rate.

Visual Examination A qualitative observation of physical characteristics with the unaided eye or within stipulated levels of magnification.

Vitreous Glassy or glasslike in color, composition, brittleness, and luster.

Vitreous Binder A glassy material that is added to thick-film paste to bind all the particles together. Dielectric pastes are nearly 100 percent glass materials. Conductor pastes used for inner layers contain 5 to 10 percent glass, while the outer layers are fritted and contain less than 1 percent. *See also Frit.*

Vitrification The conversion of a porous, ceramic material into a nonporous glasslike material as a result of being heated. The process reduces the porosity of the material and results in lower moisture absorption.

Void (1) A small hole or space in a localized area of a solid material. (2) An air bubble that has been entrapped in a plastic part during the molding process.

Volatile Condensible Materials Materials that suffer mass loss as a result of evaporation in thermal-vacuum conditions, such as outer space. These evaporated volatiles can then condense on critical surfaces such as mirrors, causing equipment performance degradation or malfunction. *See also Total Mass Loss.*

Volatile Memory A memory in which the data content is lost when power is no longer supplied to it. *See also Nonvolatile Memory.*

Voltage Breakdown *See Dielectric Breakdown.*

Voltage Drop Potential measured across a circuit protector under load conditions. This measurement is limited to the device only, and does not include voltage across connecting devices such as fuse clips. Unless otherwise noted, voltage is measured at 50 percent of the device's current rating.

Voltage Endurance The extended period of time before failure of an insulating material under voltage stress. Voltage endurance may range from seconds to years for the broad variety of operating and test conditions.

Voltage Gradient The voltage drop or change per unit length along a resistor or other conductive path.

Voltage Rating The maximum voltage a device can withstand and still safely interrupt overload current under short-circuit conditions.

Voltage Regulation A measure of the ability of a voltage regulator to maintain its output voltage under varying load conditions. *See also Voltage Regulator.*

Voltage Regulator A circuit or portion of a circuit that provides isola

tion between the load and the supply to be regulated so that the load voltage remains relatively independent of load current or input voltage fluctuations.

Volume Resistivity The electrical resistance between opposite faces of a 1 cm cube of insulating material. The value is calculated as the measured resistance in ohms times the area of the face in centimeters divided by the distance between the faces in centimeters, and is expressed as ohm-centimeters. The recommended test is American Society for Testing and Materials (ASTM) D257-54T. *See also Sheet Resistivity, Surface Resistivity.*

Vulcanization A chemical reaction in which the physical properties of an elastomer are changed by causing it to react with sulfur or other cross-linking agents. *See also Elastomer, Rubber.*

Vulcanized Fiber A cellulosic material that has been partially gelatinized by action of a chemical, usually zinc chloride, then heavily compressed or rolled to a required thickness, leached free from the gelatinizing agent, and dried.

Vulnerability The susceptibility of a circuit or system to damage under electrical or environmental stresses. This is an undesirable condition in which disturbances can damage a circuit or system.

Wafer A thin slice or flat disk, totally consisting of either a semiconductor material or a semiconductor material deposited on some substrate, in which one or more circuits or devices are simultaneously processed and subsequently may be separated into chips. *See also Semiconductor Chip.*

Wafer Scale Integration That level of integrated circuit complexity in which the complete circuit is interconnected on the wafer.

Wafer Under Test A wafer that is undergoing complex electrical testing.

Waffle Pack A flat package containing rectangular cavities for the storage and protection of bare chip devices. It resembles an egg crate and has a cover with locking devices an the sides. *Also called matrix tray.*

Warp The fibers that run lengthwise in a woven fabric. *See also Weft.*

Warpage Dimensional distortion in a plastic object after molding.

Warp and Woof The threads or wires in a woven screen that cross over and under one another at right angles.

Water Absorption The ratio of the weight of water absorbed by a material to the weight of the dry material.

Water-Extended Polyester A casting formulation in which water is suspended in the polyester resin.

Wave Guide A tube used to transmit microwave frequency electromagnetic energy.

Wave Soldering A process in which many potential solder joints are brought in contact with a wave of flowing and circulating solder for a short period of time and the joints are soldered simultaneously. *See also Soldering.*

Wear-Out (1) That span of time beyond the failure rate period in which the failure rate of the component exceeds specific predicted values. (2) The end of the useful life of a component, device, or system.

Weave The pattern in which a fabric is woven. There are standard patterns, usually designated by a style number.

Weave Exposure A surface condition of a base material in which the unbroken woven-glass cloth is not uniformly covered by resin.

Weave Texture A surface condition in which the unbroken fibers are completely covered with resin but exhibit the definite weave pattern of the glass cloth.

Wedge Bond A wire bond made with a wedge-shaped tool. Most ultrasonic bonds are wedge bonds.

Weft The fibers that run perpendicular to the warp fibers. *Also called fill or woof. See also Warp.*

Welding A process in which two metals are joined by the application of heat, causing the metals to melt and fuse together. Welding can be accomplished with or without a filler material.

Wet Lay-Up A reinforced plastic structure made by applying a liquid resin to an in-place woven or mat fabric.

Wet Strength The strength of an adhesive joint determined immediately after removal from a liquid, usually water, in which it has been immersed under specified conditions of time, temperature, and pressure.

Wettability The degree to which surface wetting occurs. Contact can be made between the solder and the metal to be soldered. *See also Wetting.*

Wetting (1) The ability to adhere to a surface immediately upon contact. (2) In soldering, the ability of molten solder to spread over a metal surface after the application of a flux and the proper amount of heat.

Whisker A metallic growth, needle-like in size and shape, that appears on the surface of a printed board or other substrate and is caused by components whose leads have been electroplated with tin and subjected to moisture, and some voltage potential. *See also Dendritic Growth.*

Wicking (1) A method of desoldering component leads or wires from a metallized hole. A prefluxed braid of stranded wire is placed on the solder joint and heated with the tip of a soldering iron. The solder is removed from the hole by capillary action wicking into the braided wire. (2) The undesirable flow of solder up along a terminal or over the surface of each conductor of a stranded wire. This results in a poorly soldered terminal, especially in J leads, or a brittle stranded wire. *See also J Lead, Solder Wicking.*

Window A holed formed by etching through an oxide or insulating layer on a semiconductor for the purpose of diffusion or deposition onto a selected area of the semiconductor.

Wiping Action The sliding action that occurs when contacts are engaged in a connector. It has the effect of removing small amounts of oxides from the contact surfaces, thereby improving conductivity.

Wirability The degree to which a package, such as a printed wiring assembly, allows the interconnection of components, hybrid packages, and terminals. The wirability of a package is optimum when there exists additional wiring capacity, via availability, and access to terminals.

Wire A solid conductor or a stranded group of metal conductors. It can be round, square, or rectangular, either bare or insulated, and have a low or relatively high resistance to the flow of current. *See also Stranded Wire.*

Wire Bond A wire connection between a pad on a semiconductor chip and a pad on a substrate. It consists of pads on both the chip and the substrate, the fine wire, and the interfaces between the wire and metal surfaces on the chip and substrate. *See also Thermocompression Bonding, Ultrasonic Bonding.*

Wire Bonding A method used to attach a fine wire, usually 1 mil in diameter, to pads on substrates and other bonding surfaces, such as semiconductors and package leads, in order to interconnect them. Methods presently used in wire bonding include thermocompression and ultrasonic and thermosonic techniques.

Wire Sag The failure of bonding wire to maintain the loop defined by the path of the bonding tool between bonds.

Wire-Wrapped Connection *See Solderless Wrapped Connection.* See Fig. 21.

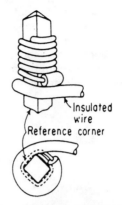

Figure 21: Wire-Wrapped Connection

Wiring A manual or automatic process of interconnecting components or chips with wires.

Wiring Overflow *See Overflow Wire.*

Woof *See Warp, Weft.*

Working Life *See Pot Life.*

Woven Fabric The flat sheet formed by interwinding yarns, fibers, or filaments. Some standard fabric patterns are plain, satin, and leno.

Woven Screen In screen printing, a screen having a specific mesh size. It may be made from stainless steel, silk, or nylon. *See also Screen Printing.*

Woven Roving A heavy glass-fiber fabric made by the weaving of roving.

Wrapped Connection *See Solderless Wrapped Connection.*

Wrist Strap A skin contact cuff that circles the entire wrist, similar to a watchband. It is made of a conductive material that is connected to ground through a current-limiting resistor so that any static charges from the wearer are rapidly and safely dissipated to ground, thus preventing any electrostatic discharge damage. *Also called personal ground strap.*

X

X-Ray Diffractometer An instrument for measuring and recording X-ray diffraction patterns and crystals, thereby permitting dialysis of their structure.

X Rays Electromagnetic radiation in the wavelength range of 0.1 to 100 angstroms produced when the inner ring of electrons (from especially heavy atoms) that have been excited by collision with a stream of fast electrons return to their ground state, thereupon giving up the energy previously imparted to them. Because of their high energy, X rays have very high ionizing and penetrating power, and can thus can be used for many types of inspection and diagnosis. *See also Electromagnetic Radiation.*

X-Y **Axis** (1) The directions parallel to the length and width dimensions of a substrate —i.e., the directions perpendicular to the Z axis. (2) The directions parallel to the fibers in a woven fiber-reinforced laminate. The thermal expansion is much lower in the *X-Y* axis, since this expansion is more controlled by the fabric in the laminate. *See also Coefficient of Thermal Expansion, Z Axis.*

Y

Yarn A combination of twisted filaments, fibers, or strands to form a continuous length suitable for weaving. *See also Weave.*

Yield Strength The minimum stress at which a material will start to physically deform without further increase in load.

Yield Value (1) The lowest stress at which a material undergoes plastic deformation. Below this stress, the material is elastic; above it, viscous. (2) The stress at which a material exhibits a specified limiting deviation from the proportionality of stress to strain. *See also Yield Strength.*

Young's Modulus The ratio of normal stress to corresponding strain for tensile or compressive stresses at less than the proportional limit of the material.

Z

Z Axis (1) The direction through the thickness of a substrate, a feature especially important for printed wiring board laminates, since thermal expansion in the Z axis is much higher than in the X-Z axis. This is because the resin in the laminate controls the Z axis thermal expansion, whereas the fabric in the laminate controls the X-Y axis thermal expansion. Resins have much high thermal expansions than do fabrics. (2) The direction perpendicular to the fibers in a woven fiber-reinforced laminate — namely, through the thickness of the laminate. Thermal expansion is much higher in the Z axis, since this expansion is more controlled by the resin in the laminate. *See also Coefficient of Thermal Expansion, X-Y Axis.*

Zener Diode A two-layer semiconductor diode designed to be operated in the reverse-bias breakdown condition.

Zero Insertion Force (ZIF) Connector A connector in which all the electrical contacts touch simultaneously, with no insertion force required, but only after the connector halves are engaged and properly aligned.

Zigzag In-Line Package (ZIP) A variation of the single in-line package (SIP), in which leads on a 0.005 inch (1.27 mm) pitch are formed to alternate sides, forming two in-line rows with a 0.10 inch (2.54 mm) pitch. *See also Single In-Line Package.*

ACRONYMS, SYMBOLS AND ABBREVIATIONS

A	angstrom unit
ABS	acrylonitrile-butadiene-styrene
AC	alternating current
AC/DC	alternating current or direct current
ACPI	automated component placement and insertion
ACS	American Ceramics Society
A/D, D/A	analog-to-digital, digital-to-analog conversion
ADC	analog-to-digital converter
ADP	automatic data processing
AES	auger electron spectroscopy
AF	audio frequency
AFC	automatic frequency control
AgPd	silver palladium
AI	artificial intelligence
AIA	Aerospace Industries Association of America, Inc.; Aircraft Industries Association of America, Inc.
AlAs	aluminum arsenide
AID	automatic insertion dip
AlGaAs	aluminum gallium arsenide
AlN	aluminum nitride
Al$_2$O$_3$	aluminum oxide (alumina)
AlSb	aluminum antimonide
ALU	arithmetic logic unit
AM/FM	amplitude modulation or frequency modulation
ANSI	American National Standards Institute
AOI	automated optical inspection
AP	arithmetic processor
APC	array processor controller
APE	asynchronous processing element
APIO	array processor input/output
AQL	acceptable quality level
ARINC	Aviation Research Incorporated
As	arsenic
ASIC	application-specific integrated circuit

ASM	American Society for Materials
ASP	advanced signed processor
ASTAP	advanced statistical analysis program
ASTM	American Society for Testing and Materials
ASW	antisubmarine warfare
ATAB	area-array tape-automatic bonding
ATE	automated test equipment; automatic test equipment
ATH	alumina trihydrate
ATR	aviation transport racking (standardized enclosures for airborne electronic assemblies)
Au	gold
AuGe	gold-germanium
AuPt	gold-platinum
AuSi	gold-silicon
AuSn	gold-tin
AV	audiovisual
AWG	American wire gauge

B	boron
Ba	barium
$BaTiO_3$	barium titanate
BCB	benzocyclobutene
BCCD	bulk channel charge-coupled device
BCW	bare chip and wire
Be	beryllium
BeO	beryllium oxide; beryllia (bromellite)
BEOL	back end of line
BiCFET	bipolar inversion-channel field-effect transistor
BiCMOS	bipolar and complementary metal-oxide semiconductor
BiCMOS-II	1.2 micron BiCMOS
BiCMOS-III	1.0 micron BiCMOS
BiCMOS-IV	0.8 micron BiCMOS
BiCMOS-V	0.5 micron BiCMOS
BiFET	bipolar field effect transistor
BiMOS	bipolar metal-oxide semiconductor
BIT	built-in test; binary digit
BITE	built-in test equipment
BLB	beam lead bonding
BLM	ball-limiting metallurgy
BMC	bulk-molding compound
BN	boron nitride
BOPS	billion operations per second
BSM	backside metallurgy

218

BTAB	bumped tape-automated bonding
BTU	British thermal unit
BW	bandwidth
BWO	backward wave oscillator
C3I	command, control, communications, and intelligence (system)
C4	controlled collapse chip connection
C	capacitance; capacitor; carbon; centigrade; graphite
CAD	computer-aided design
CAE	computer-aided engineering
CAI	computer-aided instruction
CALDAC	calibration digital-to-analog converter
CALS	computer-aided acquisition and logistics support
CAM	computer-aided manufacturing; content-addressable memory
CAPDAC	capacitor digital-to-analog converter
CAT	computer-aided testing
CAVP	complex arithmetic vector processor
CB	circuit breaker; citizen's band
CBA	chemical blowing agent
CBD	*Commerce Business Daily*
CBW	chemical-biological warfare
CC	central computer; chip carrier; control computer
CCB	controlled-collapse bonding
CCC	ceramic chip carrier
CCCC (C4)	controlled collapsible chip connection
CCD	charge-coupled device
CDA	clean dry air
CDF	cumulative distribution function
CDR	critical design review
CDRL	contract data review list
CE	concurrent engineering
CERDIP	ceramic dual in-line package (pressed ceramic/glass sealed)
CERQUAD	ceramic quad flat pack (pressed ceramic/glass sealed)
CEV	corona extinction voltage
CFC	chlorofluorocarbon; chlorinated fluorcarbon
CGA	configurable gate array
CGS	centimeter-gram-second
CIC	copper invar copper

CIGRÉ	Conférence Internationale des Grands Réseaux Électriques à Haute Tension (International Conference on High-Voltage Electrical Systems)
CIM	computer integrated manufacturing
CIV	corona inception voltage
CLA	centerline average (a measure of substrate surface roughness)
CLCC	ceramic leaded chip carrier (laminated cofired ceramic); ceramic leadless chip carrier (preferred)
CLDCC	ceramic leaded chip carrier (laminated cofired ceramic) (preferred)
CLE	coefficient of linear expansion
CM	centimeter
CMC	copper-molybdenium-copper
CML	current mode logic
CMOS	complementary metal-oxide semiconductor
CMT	chip-mounting technology
CNC	computer numerical control
COB	chip on board
COF	chip on flex
COHO	coherent oscillator
COM	computer-output microfilm
CP	centipoise; central processor
CPGA	ceramic pin grid array (laminated cofired ceramic)
CPI	cycles per instruction
CPU	central processing unit
CQFP	ceramic quad flat pack (pressed ceramic/glass sealed)
CQP	ceramic quad pack (laminated cofired ceramic)
CRIM	conventional reaction injection molding
CROM	control and read-only memory
CRT	cathode-ray tube
CSA	Canadian Standards Associations
CSIC	customer-specified integrated circuits
CSP	common signal processor
CTE	coefficient of thermal expansion
CTFE	chlorotrifluoroethylene
CVD	chemical-vapor deposition
C and W	chip and wire
D/A	digital to analog
DAC	digital-to-analog converter
DAIP	diallyl isophthalate
DAM	data-addressable memory
DAP	diallyl phthalate

220

DASD	direct-access storage device
dB	decibel
DC	direct current
DCA	direct chip attach
DCAS	Defense Contracts Administration Services
DESC	Defense Electronics Supply Center
DFM	design for Manufacturability
DFT	design for test
DGEBA	diglycidyl ether of bisphenol A
DI	deionized water
DIN	Deutsches Institute für Normung (German Standards Organization)
DIP	dual in-line package
DMA	direct memory access; dynamic mechanical analysis
DMM	digital multimeter
DNP	distance to neutral point
DOD	Department of Defense (U.S.)
DODISS	Department of Defense Index of Specifications and Standards
DOE	Department of Energy (U.S.)
DOP	dioctyl phthalate
DOS	disk-operating system
DPA	destructive physical analysis
DPDT	double pole double throw
DPST	double pole single throw
DPU	data processor unit
DRAM	dynamic random-access memory
DSC	differential scanning calorimetry
DSO	digital storage oscilloscope
DSP	digital signal processor
DTA	differential thermal analysis
DTAB	demountable tape-automated bonding
DTIC	Defense Technical Information Center
DTL	diode-transistor logic
DUT	device under test
DVM	digital voltmeter
DWF	dice in wafer form
E	voltage; potential
EAPROM	electrically alterable programmable read-only memory
EC	engineering change; European Community (Belgium, Denmark, France, Germany, Greece, Ireland, Italy, Luxembourg, Netherlands, Portugal, Spain, United Kingdom)

ECCM	electronic counter-countermeasure
ECL	emitter-coupled logic
ECM	electronic countermeasure
ECO	electron-coupled oscillator
ECTFE	ethylenechlorotrifluoroethylene
EDM	electrical discharge machine
EDP	electronic data processing
EDS	energy-dispersive spectroscopy
EEPROM	electrically erasable programmable read-only memory
EHF	extremely high frequency
EIA	Electronics Industries Association
EIAJ	Electronics Industry Association of Japan
EL	electroluminescense
ELF	extremely low frequency
EMC	electromagnetic compatibility
EMF	electromagnetic force
EMI	electromagnetic interference
EMP	electromagnetic potential; electromagnetic pulse
EMPF	electronic manufacturing productivity facility
EMR	electromagnetic radiation
ENOB	effective number of bits
ENR	excess noise ratio
EO	electrooptic
EOL	end of life
EOS	electrical overstress
EOSP	electrooptic signal processor
EOSPC	electrooptic signal processor controller
EP	epoxy
EPA	Environmental Protection Agency
EPIC	environmentally protected integrated circuit (laminated plastic chip carrier)
EPLD	erasable programmable logic device
EPR	ethylene propylene rubber
EPROM	electrically programmable read-only memory; erasable programmable read-only memory
ESCA	electron spectroscopy for chemical analysis
ESD	electrostatic discharge
ESDS	electrostatic discharge sensitive
ETFE	ethylenetetrafluoroethylene
ETPC	electrolytic tough pitch copper
ETV	educational TV
EUT	equipment under test
EVA	ethylene vinyl acetate
EW	electronic warfare

f	farad; frequency
FA	failure analysis
FACI	first article configuration inspection
FAR	federal acquisition regulations
FCB	flip chip bonding
FCC	flat conductor cable; flexible conductor cable
FCFC	flat conductor flat cable
FCT	field-controlled thyristor
FDDI	fiber-distributed data interface
FEA	finite element analysis
FEOL	front end of line
FEP	fluorethylene propylene; fluorinated ethylene propylene
FET	field-effect transistor
FIC	film-integrated circuit
FLIR	forward-looking infrared
FM	frequency modulation
FMEA	failure mode and effects analysis
FMECA	failure mode, effects, and criticality analysis
FO	fiber optics
FP	fine pitch; flat pack
FPAP	floating point arithmetic processor
FPC	fine-pitch chip carrier
FPGA	field-programmable gate array
FPT	fine-pitch technology
FPW	flexible printed wiring
FQFP	fine-pitch quad flat pack (molded plastic)
FR	failure rate; fiber reinforced; flame retardant; flammability rating
FR-4	flame-retardant epoxy glass laminate; designated by National Electronic Manufacturers Association
FRP	fiber reinforced plastic
FRU	field-replaceable unit
FS	full scale
FSR	full-scale range
FTR	functional throughput rate

G	giga (10^9); gravitational force
Ga	gallium
GaAs	gallium arsenide
GaAsP	gallium arsenide phosphide
GaInAs	gallium indium arsenide
GaP	gallium phosphide
GaSb	gallium antimonide
GCA	ground-controlled approach
GDT	geometric dimensioning and tolerancing

Ge	germanium
GFE	government-furnished equipment
gHz	gigahertz (10^9)
gnd	ground
GPC	general-purpose computer
GSA	General Services Administration
GTAB	ground plane tape-automated bonding
GW	gull wing (lead)
H_2	hydrogen molecule
HAZMAT	hazardous material
HCC	hermetic chip carrier
HCMOS	high-density complementary metal-oxide semiconductor
HDCM	high-density ceramic module
HDI	high-density interconnections
HDMI	high-density microelectronics interconnections
HDPE	high-density polyethylene
HDT	heat deflection temperature; heat distortion temperature
HDTV	high-definition television
HEMP	high altitude electromagnetic pulse
HEMT	high-energy mobility transistor
HF	high frequency
HFET	heterostructure field-effect transistor
HIC	hybrid integrated circuit
Hi-K	high-dielectric constant
HIPS	high-impact polystyrene
HMIC	hybrid microwave integrated circuit
HMOS	high-performance metal-oxide semiconductor
H_2O	water
H_2O_2	hydrogen peroxide
HOL	higher-order language
HPM	high-power microwaves
HTRB	high-temperature reverse bias
Hz	hertz
i	current
IACS	international annealed copper standard
IAPU	image array processing unit
IC	integrated circuit; internal connection
IDC	insulation displacement connector; insulation displacement contact
IEC	International Electrotechnical Commission

IEEE	Institute of Electrical and Electronics Engineers
IEEJ	Institute of Electrical Engineers of Japan
IEPS	International Electronics Packaging Society
IF	intermediate frequency
IGBT	insulated gate bipolar transistor
IGFET	insulated gate field-effect transistor
ILB	inner layer board; inner lead bond (er) (ing)
IMC	insertion-mount component
IMS	insulated metal substrate
IMST	insulated metal substrate technology
IMT	insertion-mount technology
In	indium
InP	indium phosphide
InSb	indium antimonide
I/O	input/output
IPC	Institute for Interconnecting and Packaging Electronic Circuits (formerly Institute of Printed Circuits)
IPN	interpenetrating polymer network
IPS	instructions per second
IR	infrared; insulation resistance
IRED	infrared-emitting diode
IRS	infrared scan
ISA	imaging sensor autoprocessor; instruction set architecture
ISHM	International Society for Hybrid Microelectronics
I/SMT	interconnect/surface-mount technology
ISO	International Standards Organization
ITV	industrial television

JAN	joint Army-Navy
JC	JEDEC committee
JEDEC	Joint Electronic Devices Engineering Council
JFET	junction field-effect transistor
JIT	just-in-time

K	dielectric constant; kelvin; thousand (10^3)
KGD	known good die
kHz	kilohertz (10^3 Hz)

L	inductance
LAN	local area network

LC	inductance-capacitance
LCC	leadless chip carrier
LCCC	leadless ceramic chip carrier (laminated cofired ceramic)
LCD	liquid crystal display
LCP	liquid crystal polymer
LD	laser diode
LDCC	leaded ceramic chip carrier (laminated cofired ceramic)
LDPE	low-density polyethylene
LED	light-emitting diode
LEMP	lightning electromagnetic pulse
LF	low frequency
LGA	land grid array
LHD	leadless hybrid device
LIC	linear integrated circuit
LID	leadless inverted device
LIM	liquid injection molding
LLDPE	linear low-density polyethylene
LMCH	leadless multiple chip hybrid
LNA	low-noise amplifier
LO	local oscillator
LOI	limiting oxygen index
LPCVD	low-pressure chemical vapor deposition
LPE	liquid-phase epitaxy
LPGA	leadless pad grid array
LPPQFP	low-profile plastic quad flat pack
LRM	line-replaceable module
LRU	line-replaceable unit
LSB	least significant bit
LSI	large-scale integration
LST	logic surface terminal
LTPD	lot tolerance percent defective

M	million (mega)
m	meter
mm	millimeter
MANTECH	manufacturing technology program
MBT	metal base transistor
MC	metallized ceramic
MCC	Microelectronics and Computer Technology Corporation; miniature chip carrier; multiple chip carrier
MCM	multichip module
MCM-C	multichip module (ceramic substrate, either cofired or low-dielectric-constant (K) ceramic)

MCM-D	multichip module (with deposited wiring on silicon substrate or ceramic and metal substrate)
MCM-L	multichip module (high-density, laminated printed circuit boards)
MCP	multichip package
MCRPQFP	molded carrier ring plastic quad flat pack
MEK	methyl ethyl ketone
MEKP	methyl ethyl ketone peroxide
MELF	metal electrode face (bonded)
MESFET	metal-semiconductor field-effect transistor
MF	medium frequency
MFD	microelectronic functional device; microfarad
MFLOPS	million floating-point operations per second
MHP	multichip hybrid package
mHz	megahertz (10^6 Hz)
MIC	microwave integrated circuit; monolithic integrated circuit
micrometer	micron (10^{-6} meter)
micron	micrometer (10^{-6} meter)
mil	one-thousandth of an inch (0.001 inch)
MIL-STD	military standard
MIPS	million instructions per second
MIS	metal-insulator-semiconductor
MKS	meter-kilogram-second
MLB	multilayer board
MLC	multilayer ceramic (laminated cofired ceramic)
MMIC	monolithic microwave integrated circuit
MMPQFP	multilayer molded plastic quad flat pack
MMU	memory-management unit
MNO$_2$	manganese dioxide
MNOS	metal nitride-oxide semiconductor
MNS	metal-nitride semiconductor
MOPS	million operations per second
MOS	metal-oxide semiconductor
MOSFET	metal-oxide-semiconductor field-effect transistor
MOST	metal-oxide-semiconductor technology
MPU	microprocessing unit
MRB	material review board
MS	microstrip
MSB	most significant bit
MSI	medium-scale integration
MTBF	mean time between failures; mean time between faults
MTM	multiple termination module
MTNS	metal thick-nitride semiconductor
MTOS	metal thick-oxide semiconductor
MTTF	mean time to failure

MTTR	mean time to repair
MVT	moisture vapor transmission
MW	molecular weight
MWD	molecular weight distribution

n	nano (billionth)
N$_2$	nitrogen molecule
NA	numerical aperture
NASA	National Aeronautics and Space Administration
NATA	North American Telecommunications Association
NBS	*See NIST*
NC	normally closed; numerically controlled
NDE	nondestructive evaluation
NEAT	nothing else added to it
NEMA	National Electrical Manufacturers Association
NEMP	nuclear electromagnetic pulse
NEP	nuclear electromagnetic pulse
NiCd	nickel cadmium
NIRA	near infrared reflectance analysis
NIST	National Institute of Standards and Technology (formerly National Bureau of Standards, NBS)
NMOS	N-channel metal-oxide silicon
NO	normally operative
NO$_2$	nitrogen peroxide
NO$_3$	nitrogen trioxide
N$_2$O$_2$	nitrous oxide
NR	natural rubber
NRE	nonrecurring engineering
ns	nanosecond (10^{-9} or one-billionth of a second)

O$_2$	oxygen molecule
O$_3$	ozone molecule
OA	organic acid
OEIC	optoelectronic integrated circuit
OFHC	oxygen-free high-conductivity (copper)
OH$^-$	hydroxyl ion
OLB	outer lead bond (er) (ing) (tape-automated bonding)
OMC	organic matrix composite
OPS	operations per second
OS	operating system
OSHA	Occupational Safety and Health Act

P	phosphorous
p	pico (trillionth)
PA	polyamide (nylon)
PAA	pad area array; pin grid array
PAI	polyamide imide
PAL	programmable array logic
PAN	polyacrylonitrile
Pa.s	Pascal second
PBT	polybutylene terephthalate
PC	personal computer; polycarbonate; printed circuit
PCB	printed circuit board
PCBT	pressure cooker bias test
PCC	*See PLCC*
PCR	plastic chip carrier, rectangular
PCT	pressure cooker test
PCTFE	polychlorotrifluorethylene
PDA	percent defective allowable
PDID	plastic dual in-line package (molded plastic)
PDIP	plastic dual in-line package (molded plastic)
PEEK	polyetherether ketone
PEL	picture element in display; pixel
PES	porcelain-enamel-steel
PET	polyethylene terephthlate; porcelain-enamel technology
PF	power factor
PGA	pad grid array; pin grid array (laminated cofired ceramic)
PHP	power hybrid package
PHR	parts per hundred parts resin
PI	polyimide
pi	3.1416
P/I	packaging and interconnecting
PIC	photonic integrated circuit
PIN	P-N junction with isolation region (diode)
PIND	particle impact noise detection (testing)
PIP	pin insertion package
PLA	programmable logic array
PLCC	plastic leaded chip carrier (molded plastic)
PLD	programmable logic device
PLDCC	plastic leaded chip carrier (preferred)
PLOS	P-channel metal-oxide-semiconductor
PMMA	polymethyl methacrylate
PMP	premolded plastic
POS	porcelain-on-steel (substrate)
PP	polypropylene
PPGA	plastic pin grid array (laminated plastic)
PPM	parts per million; pulse-position modulation

PPO	polyphenylene oxide
PPP	parallel pipeline processor
PPS	polyphenylene sulfide
PQFP	plastic quad flat pack (molded plastic); standardized by JEDEC
PROM	programmable read-only memory
PS	polystyrene
ps	picosecond (10^{-12} or one trillionth)
PSCR	photosensitive copper reduction process
PSG	phosphosilicate glass
PSI	pounds per square inch
PSIG	pounds per square inch gauge
PSP	programmable signal processor
Pt	platinum
PTF	polymer thick film; precision thin film
PTFE	polytetrafluoroethylene
PTH	pin through hole; plated through hole
PTM	pulse time modulation
PUR	polyurethane
PVA	polyvinylalcohol
PVC	polyvinyl chloride
PVD	physical vapor deposition
PVF	polyvinyl fluoride
PWA	printed wiring assembly
PWB	printed wiring board
PWM	pulse-width modulation
PX	paraxylylene

QCI	qualification conformance inspection
QFB	quad flat butt-leaded package
QFB/EIAJ	quad flat butt-leaded package/Electronics Industry Association of Japan
QFP	quad flat pack
QML	qualified manufacturers list
QPL	qualified products list
QUIP	quad in-line package (molded plastic)

R	rosin flux
r	resistance; resistor
RA	rosin activated flux
RAM	random-access memory
RC	resistance capacitance
RCC	rectangular chip carrier

RF	radio frequency
RFI	radio frequency interference
RGA	residual-gas analysis
RH	relative humidity
RI	receiving inspection
RIE	reactive-ion etching
RIM	reaction injection molding
RISC	reduced-instruction-set computer
RISP	rapid impingement speed plating
RLC	resistance-inductance capacitance
RMA	rosin mildly activated
rms	root mean square
ROM	read-only memory
RP	reinforced plastic
RPM	revolutions per minute
RPV	remotely piloted vehicle
RTI	radiation transfer index
RTL	resistor-transistor logic
RTM	resin-transfer molding
RTP	reinforced thermoplastic
RTV	room-temperature vulcanizing

SAM	scanning acoustic microscope; scanning auger microscope; surface-to-air missile
SAMPE	Society for the Advancement of Material and Process Engineering
SAW	surface acoustic wave
Sb	antimony
SB-DIP	side-brazed dual in-line package (laminated cofired ceramic)
SCC	square chip carrier; stress-corrosion cracking
SCD	source control drawing; specification control drawing
SCE	storage control element
SCM	single chip module; *same as SCP*
SCOT	sealed chips on tape
SCP	single chip package; *same as SCM*
SDI	strategic defense initiative
SDP	static discharge pulse
SE	shielding effectiveness
SEM	scanning electron microscope; standard electronic module
SFI	solder free interconnections
SHF	super high frequency
Si	silicon
SiC	silicon carbide

SiN	silicon nitride
SiO	silicon monoxide
SiO₂	fused silicon (silica glass); quartz (crystalline silicon); silicon dioxide (silica)
SI	silicone; Système International (international systems of units)
SIM	single in-line module (laminated cofired ceramic and laminated plastic)
SIMM	single in-line memory (laminated cofired ceramic and laminated plastic)
SIMS	secondary ion mass spectrometry
SIP	single in-line package (laminated cofired ceramic and laminated plastic)
SIR	surface insulation resistance
SLAM	scanning laser acoustic microscope; single-layer alumina metallized
SLC	single-layer ceramic
SLT	solid logic technology
SM	surface mount
SMA	surface-mounted assembly
SMART	surface mount and reflow technology
SMC	sheet-molding compound; surface-mount component
SMD	standard military drawing; surface-mount device
SMOBC	solder mask over bare copper
SMP	surface-mount package
SMPGA	surface-mount pin grid array (laminated cofired ceramic)
SMT	surface mount technology
Sn	tin
SO	small outline (package with gull wing leads, molded plastic)
SOB	small outline butt-leaded package
SOI	silicon on insulator
SOIC	small outline integrated circuit (package with gull wing leads, molded plastic)
SOJ	small outline J-leaded package, molded plastic
SOP	small outline package, molded plastic; *also called SOIC*
SOS	silicon on sapphire
SOT	small outline transistor
SOW	statement of work
SP	signal processor
SPC	statistical process control
SPDT	single pole double throw
SPE	Society of Plastics Engineers
SPI	Society of the Plastics Industry

SPICE	simulation program for integrated circuits emphasis
SPS	systolic processing superchip
SPST	single pole single throw
SQC	statistical quality control
SQFP	shrink quad flat package
SRAM	short range attack missile; static random-access memory
SRC/MCC	Semiconductor Research Corporation/Microelectronics and Computer Technology Corporation
SREMP	source region electromagnetic pulse
SRU	smallest replaceable unit
SSI	small-scale integration
SSOIC	shrink small outline integrated circuit
SSOP	shrink small outline package
SST	solid state technology; self-stretching solder technology
SSWS	static-safe work station
STL	stripline

Ta	tantalum
Ta$_2$O$_5$	tantalum pentoxide
TAB	tape-automated bonding
TC	temperature coefficient; thermocompression (bonding)
TCC	temperature coefficient of capacitance
TCE	temperature coefficient of expansion; trichloroethylene
TCM	thermal conduction module
TCP	tape carrier package
TCR	tape carrier ring; temperature coefficient of resistance
TEOS	tetraethoxysilane
TF	trim and form
TFE	tetrafluoroethylene
TFSOI	thin-film small outline integrated circuit package
TFT	thin-film transistor
Tg	glass transition temperature
TGA	thermogravimetric analysis
TH	through-hole mount
T/H	track and hold
THB	temperature humidity bias
Ti	titanium
TIC	tape-automated bond in cap
TiO$_2$	titanium dioxide (titania)

TiSi$_2$	titanium silicide
TLC	thin-layer chromatography
TMA	thermomechanical analysis
TO	transistor outline package
TOP	thin-outline package
TPE	thermoplastic elastomer
TQFP	tape quad flat pack
TQM	total quality management
TREE	transient radiation effects on electronics
TS	thermoseting; thermosonic (bonding)
TSM	thermosetting material; top-side metallurgy
TSOP	thin small outline package
TSSOP	thin scaled small outline package
TTL	transistor-transistor logic
TVI	TV interference
TWA	traveling-wave amplifier
TWT	traveling-wave tube

UFPT	ultra-fine-pitch technology
UHF	ultra-high frequency
UL	Underwriters Laboratory
ULSI	ultra-large-scale integration
US	ultrasonic (bonding)
UUT	unit under test
UV	ultraviolet

v	volt
VCD	variable center distance
VF	voice frequency
VGA	video graphics array
VHF	very high frequency
VHSIC	very high speed integrated circuit
VIL	vertical in-line
VLF	very low frequency
VLSI	very large-scale integration
VPS	vapor phase soldering
VQFP	very small quad flat pack (molded plastic)
VSO	very small outline package
VSOIC	very small outline integrated circuit
VSOP	very small outline package (molded plastic)

w	watt
WB	wire bonded
WIP	work in progress
WPM	words per minute
WSI	wafer scale integration
WUT	wafer under test
WVT	water vapor transmission

X	reactance
X-MOS	high-speed metal-oxide-semiconductor

Y	yttrium
YAG	yttrium-aluminum-garnet
YIG	yttrium-iron-garnet

Z	impedance
ZIF	zero insertion force
ZIP	zigzag in-line package
Zn	zinc
ZnO	zinc oxide
ZnS	zinc sulfide

About the Authors

CHARLES A. HARPER is president of Technology Seminars, Inc., a leading organization serving the educational needs of the electronics industry. He is also editor in chief of the *McGraw-Hill Electronic Packaging and Interconnection Series* and of the *Electronic Packaging and Interconnection Handbook*. Mr. Harper has extensive experience in the applications and management of electronic packaging and interconnections for all types of electronic systems and assemblies. He is Past President of the International Electronics Packaging Society and a member of the International Society for Hybrid Microelectronics, IICIT, the Society for the Advancement of Material and Process Engineering, and the IEEE.

MARTIN B. MILLER is senior associate at Technology Seminars, Inc. He has more than thirty years experience in all phases of electronic assemblies, packaging, and materials, and has managed a laboratory serving both engineering and manufacturing.